AF605380

THE MAN NEXT DOOR

AND OTHER TRUE STORIES OF WAR AND PEACE

MIKE COLMAN

STOKE HILL PRESS

First published in 2023

by Stoke Hill Press
c/ 122 Wellbank Street
Concord NSW Australia 2137
www.stokehillpress.com

A catalogue record for this work is available from the National Library of Australia

ISBN: 978-0-6487331-6-4

Cover Design by Luke Causby, Blue Cork
Internal Design & Typesetting by Kirby Jones
Printed in Australia by Ligare Book Printers

Front Cover Quote
The words at the top of the front cover come from the foreword by Keith Payne VC AM.

Sources
Chapter 1, *The Man Next Door*, was written especially for this book. The other stories by Mike Colman in this collection were first published in *QWeekend*, the colour magazine in Saturday's *Courier-Mail*. The original publication dates for these stories are listed on page 238.

Back Cover
The images on the back cover — Avenge the Nurses! (Reference Number: ARTV09088), Lance Corporal Charles Blackman (RCDIG0000836), Warrant Officer Class 2 Keith Payne VC AM (LES/69/0593/VN) and Sutton Veny (H01736) — are reproduced with the kind permission of the Australian War Memorial.

This book is dedicated to Greg Growden,
Wayne Smith and Paul Malone.

CONTENTS

FRONT COVER

The remarkable wartime memoir of Sergeant Ray McMillan, the 'Man Next Door' whose story is the basis for Chapter 1 of this book, contains a black-and-white photocopy of a painting, with the following caption:

> Impression by an artist, after the war, of my Hudson being attacked by Japanese Zeros following our bombing of the Japanese invasion fleet on Sumatra in February 1942. The Hudson was piloted by F/Lt Diamond with crew of P/O Evans, Sgt Walsh and myself.

There is a signature in the bottom right-hand corner of the painting: Frank Harding. A google search couldn't locate the artwork, but it did reveal that similar paintings by Frank Harding, an acclaimed artist who died in 1990, were on display at a gallery in his old hometown of Renmark, which is located on the Murray River in South Australia, 250km northeast of Adelaide, near the Victorian border. Stoke Hill Press publisher Geoff Armstrong contacted Megan Harding, Frank's daughter, who manages the gallery, and Megan confirmed that the painting was indeed by her father. She also revealed that Frank Harding actually produced two paintings of Flight Lieutenant Ossie Diamond's Hudson. The second, 'Sinking the *Awazisan Maru*', appeared in his limited-edition book, *They Flew For The King*, published in 1990.

From the moment Armstrong first saw 'Sinking the *Awazisan Maru*', he was keen for it to be on the cover of *The Man Next Door.* Ray McMillan describes the battle in his memoir, writing at one point: 'On our return to base we were not surprised to see we had been hit in a number of places by enemy gunfire, or shrapnel caused by our bombing. One missile struck the starboard motor's oil line, and it was estimated the engine would have seized within a couple of minutes.' Megan Harding was happy to help, but there was a problem. It was a month before Christmas 2022, and the Murray River upstream was in flood. Renmark was bracing itself for a potential crisis similar to that which deluged the town in 1956. The gallery, which in normal circumstances is located less than 500 metres from the riverbank, was under threat. The precious artwork was being moved to higher ground.

Fortunately, Renmark had time to prepare, the levees largely held, and the damage in and around the town was minimised. Not everyone on the Murray floodplain was so lucky. In the new year, the gallery re-opened and designer Luke Causby was able to produce the cover that adorns this book. It is a privilege to be able to showcase Frank Harding's work, and we are extremely grateful for the support of Megan and the entire Harding family.

Frank Harding Gallery is located at 31 Murtho Street, Renmark, South Australia. For more information, ring 08 8586 6972 or go to www.murrayriver.com.au/renmark/frank-harding-gallery/

ACKNOWLEDGMENTS

As with all books, this one could not have been possible without the involvement and support of a lot of people.

Firstly, I would like to express my sincere appreciation to the family of Ray McMillan ('The Man Next Door') — his daughters Linda and Nola, and son Stephen — for permission to use their father's extraordinary memoir as the basis for the opening chapter and spiritual touchstone of this collection of stories. Ray's wartime experiences that are recalled so meticulously and at times shockingly — and the way he somehow managed to put those experiences behind him as he got on with life in the years following his return home — epitomise what this book is about. These are not just stories of war heroes like the ones I read about when I was growing up in the 1960s and '70s — although Keith Payne certainly falls into that category. They are stories about ordinary men and women placed in the pressure cooker of war; how they coped with that pressure and how they and those around them were affected by what they went through. These are the stories of just a handful, but they represent the experiences of many more.

The seeds of this book were sown in 2006 in the blood-soaked soil of Gallipoli. The *Courier-Mail* newspaper had recently begun publishing a Saturday colour magazine titled *Qweekend* that offered the paper's journalists previously unimaginable possibilities in terms of time, space and budget. It gave me, a senior sportswriter on the newspaper, an opportunity to fulfil an ambition I had harboured for years.

'How about,' I asked the magazine's editor Christine Middap, 'you send me to Gallipoli on Anzac Day so I can follow in the footsteps of my grandfather and great-uncle? I'll write a story and you hold on to it for 12 months and publish it on Anzac Day 2007?'

'Sure,' Christine answered, to my astonishment. 'Let's do it.'

That story, which appears in this collection under the title *Charlie Marshall Says G'day*, led to me joining *Qweekend* fulltime. It became something of a tradition that I would write a piece to appear every Anzac Day — stories that make up the majority of this book. One of them, *The Tree of Life*, won the Walkley Award for Best Magazine Feature in 2011 and led to the writing of my book, *Crew*.

The pilgrimage to Gallipoli took me down another path as well. On the return trip I had a long stopover in Singapore and, still on a high from the excitement and emotion of the previous three days, I headed to a bank of computers and began looking up Australian military history sites. One I logged onto covered the records of the nation's Victoria Cross recipients, and I saw that our only living VC at that time was Queenslander Keith Payne, Australia's most decorated soldier of the Vietnam War. I decided to contact him when I got home. The result was the book *Payne VC* and a long friendship with Keith and his wife Flo. It was an honour to have him write the introduction to this book.

The next person I should acknowledge is publisher Geoff Armstrong, whose Stoke Hill Press specialises in sports books, not military, although sometimes the line gets blurred. His beautifully produced hardcover, *Cecil Healy: A Biography*, by John Devitt and Larry Writer, tells the story of Australia's only Olympic gold medallist to be killed while serving on the battlefield. Geoff, who published my last book, a collection of sports stories titled *Mike Colman On Sport,* has been a great supporter of mine over the years. Perhaps he's taken a risk by getting behind this project, but as he explains in his publisher's note at the back of the book, he also has very personal reasons for doing so. Geoff's tribute to his grandfather Ted Armstrong is worthy of a chapter of its own.

Finally, I must pay tribute to my wife Linda, whose support, advice and proof-reading skills have made these stories, and many others that I have written over the past 40 years, possible.

Mike Colman

August 2023

FOREWORD

By Keith Payne VC AM

I AM VERY PLEASED to write the foreword to this collection of military-themed stories by Mike Colman. The book is a project close to my heart for many reasons. I have known Mike for over 15 years. He called me out of the blue in 2006 and asked me if he could write a book about me. My first question was, 'How much is this going to cost me?' That is how I felt back then, because I honestly couldn't believe that anyone would be too interested in my story, let alone be willing to pay to read it. In fact, after the way I and my fellow servicemen and women were treated when we returned from Vietnam more than 50 years ago, I wasn't sure what sort of reaction any book about me or that war would stir up.

When I got back to Brisbane in September 1969 after being awarded the Victoria Cross, I was the flavour of the month. A picture of me being greeted by my wife Flo and our five boys was on the front page of all the newspapers, I was handed the key to the city and we were given a week's free holiday, but when that month ended, it really ended. I came home from work at the Royal Military Academy at Duntroon in Canberra and 'Baby Killer' had been painted on my front fence. My sons virtually had to fight their way through school because their father had been to Vietnam and our superiors told us to always wear civvies when we went into town to avoid trouble. It is a sad thing when you can't wear your country's uniform on the streets of Australia.

I'm glad to say attitudes changed with time. One of the proudest experiences of my life was taking part in the Welcome Home parade in Sydney on Anzac Day 1987, when 22,000 Vietnam veterans were finally thanked and embraced by 100,000 of their countrymen and women. That, to me, marked a turning point. From that day on it became okay to recognise the sacrifices that people in uniform and their families have made and continue to make for our country.

No one is glorifying war, least of all people who have experienced it, but it is still important to tell the human stories about those who have served, and those they left behind.

That is what makes books like this one so valuable. It isn't a novel or a John Wayne movie script. It is a collection of real stories about real people and how they were affected by their wartime experiences. No one who has been through war, or whose loved one has, can be unaffected. So many soldiers come back from serving overseas suffering Post Traumatic Stress Disorder. I did myself when I came back from Vietnam, so I know what they're going through.

In recent times I've gone over and visited the troops in Afghanistan and Iraq. I look into their eyes and they've got the 'nine mile stare' and I think, *You poor bugger.* I know they've got a hard road ahead of them. For them to feel that people are behind them and appreciate what they have done means a lot.

During the many hours that Mike and I spent together when he was working on my book *Payne VC,* we got to know each other well.

He told me that he was inspired to write about me after going to Gallipoli in 2006 to follow in the footsteps of his grandfather and great-uncle, who both went ashore at Anzac Cove under heavy fire on April 25, 1915. The story of his trip is in this collection.

In 2010, Mike was all set to accompany me and my former US Special Forces medic, Jerry Dellwo, back to Vietnam as we searched for the remains of our fallen comrade, Monty Montez. The local police chief refused to let Mike into the province at the last moment, but he was there waiting for me at the airport when I got back. That story is in the book too.

There are plenty of others: about war brides and the Rats of Tobruk and the fight for recognition for World War I Aboriginal Digger Charlie Blackman amongst others. And there is the shocking but inspiring story of Ray McMillan, 'The Man Next Door', and his suffering at the hands of the Japanese.

Every story is different, but they have a common thread running through them. As the cover says, they are all true stories about war and peace. *The Man Next Door* is a good book that deserves to be read.

Chapter 1

THE MAN NEXT DOOR

HE WAS THE MAN NEXT DOOR. His name was Ray, but I knew him only as Mr McMillan.

It was a very ordinary suburban street on Sydney's North Shore. We moved in in 1957 when I was nearly two years old.

The McMillans were already there when we arrived, and so were the Suttons in the next house up and the Donalds across the road. Everyone had kids around the same age, so we all knew each other pretty well growing up. Michael Sutton was my best friend. When we were about 10 years old, he told me that his dad had been a sergeant in the war. Another time, Bruce Donald from across the road and I were playing underneath his house and came across an old tin trunk of his father's that had the words 'Lt I. Donald' and some numbers painted in neat white letters along the side. It was no big deal in those days. Nearly everybody's dad had been in the war.

My father's best mate, Bruce Langford, lived a couple of streets away. Bruce was the local bank manager and Dad first met him when he went in to apply for a home loan. Bruce had been a captain in Borneo and brought back a Samurai sword that he used to slash the long grass at the back of their yard. Another family friend was Marshall Burgess, a Qantas pilot who had flown Catalina flying boats in the war, something I didn't know until I saw an ABC TV news item about the last meeting of the Catalina Association in 2011. Sitting in the cockpit of the only Catalina still flying in Australia, Marsh, then 89, told the reporter about the night he nearly didn't make it back.

> I flew it down to about 20 feet above the water and if you can imagine, it was a pitch-black night and these bullets were whizzing everywhere around us, thousands of the things. Fortunately, they didn't hit us. How they missed us I don't know.

Like so many others, Marsh Burgess never talked about what he had gone through in the war. Not until much, much later anyway. Men like him were all around you in the 1960s and '70s. Not heroes necessarily, not old soldiers with a chest full of medals; just ordinary blokes who had gone away to war and done extraordinary things. It was many years after Bruce Langford's death that I stumbled onto a picture in the Australian War Memorial archives. The caption says: 'Beaufort, Borneo. August 27, 1945. Captain HB Langford, Officer Commanding, and members of B Company, 2/43rd Infantry Battalion, who have just returned from a patrol in the area.' Surrounded by his smiling comrades, Bruce is showing off the Samurai sword that I would later see him using in his garden. Crouching to his left, holding a Bren gun, is Private Tom Starcevich, who two months earlier had used that weapon to single-handedly clean out four Japanese machine-gun nests that had the company pinned down, earning him the Victoria Cross.

One of ten children raised on a farm at Grass Patch in Western Australia, Tom Starcevich returned to the area after the war and bought a small farm of his own. A quiet, shy man who avoided the spotlight, it was not until six years after his death in 1989 that a statue commemorating his wartime service was unveiled at Grass Patch.

That's how it was back then. You wouldn't know who had a tale to tell about what they had been through in the war. It could be the man who left the milk bottles on your front step every second morning, or the one who collected your ticket at the train station, or who drove the school bus. It could be the man next door.

RAY WAS A SINEWY, fit-looking man who always seemed to be tanned because he spent so much time in the garden. Their house stood out on our street of blond and red-brick bungalows because it was painted white, with a picket fence and a perfectly manicured front lawn.

I don't remember Ray ever touching up the paint job, but he must have, because it glowed, and the garden was immaculate. Roses out the front, fruit trees and veggies in the back. My mum was a keen gardener too, and she and Ray would sometimes talk through the back fence as they weeded or pruned. Never face-to-face, always through the fence, although occasionally they would pass over a cutting or some fruit they had grown.

Ray had been in the war too. I found that out when I was about 12 years old. For as long as I could remember my parents owned a Morris Minor, but around 1967 it was starting to show its age. Even using the crank handle on cold mornings didn't always get it started. And so, with great excitement, Dad traded in the old Moggie for a brand new Toyota Corolla. Talk about state of the art. It was turquoise with a white interior, and while it didn't have the automatic aerial of the top-of-the-range Toyota Crown, it did feature a radio and push-button cigarette lighter.

Dad brought it home on a Friday and the next morning he and I sat in it on the driveway reading the owner's manual, flicking switches and testing the horn. As we did, I noticed Ray looking at us from his front yard. Dad hopped out and walked across to him, maybe going to offer a spin around the block. Their conversation was short and appeared rather terse. Ray didn't seem to be sharing our joy about the new purchase. He shook his head, nodded grimly at the car, and headed inside. When Dad came back, I asked him what was going on.

'He was a POW in the war,' he told me. 'He said he'd never sit in a Japanese car.'

The last time I saw Ray was in February 2002, at my mother's funeral. I'd long before moved to Queensland and started a family of my own, but my parents had stayed in the old house. Ray was still next door, the Suttons remained one house up and the Donalds were still across the road. Ray walked up to me after the service and shook my hand, reminding me who he was, as if I didn't know. He seemed smaller but just as wiry and tanned as I remembered.

'I'll tell you what Michael,' he said, looking genuinely sad. 'I'm really going to miss your mum.'

It was probably the longest conversation we'd ever had.

Ray died in November 2007, aged 91. My father went to the funeral where he spoke to Ray's son Stephen, who is now a barrister in Sydney. Stephen had delivered a eulogy and mentioned that a few years before his death, his father had written a memoir recounting his wartime experiences. My father told Stephen that he felt I would be interested in reading it, and a few weeks later a photocopied document arrived in the post.

A Prisoner of War on His Journey to Japan
1939-1945

By

Raymond C McMillan

November 2005

Written by hand on the title page was the dedication:

> I present this brief account of my journey with the enemy to my son, Stephen. It was long and stressful, and the Japanese were extremely cruel and sadistic. Living was a day-to-day proposition but I always felt I would eventually make a safe return home.

I was about to learn more about the man next door. The following draws heavily on Ray's memoir.

RAY WAS BORN AT Wee Waa, NSW, on February 7, 1916, the third youngest of a family of eight children. His father, James — known as Jim — was a farmer who had owned a grazing property near Coonamble on the central western plains of NSW. When that was not successful, he purchased *Paisley*, a property 16km south of Wee Waa. From Coonamble to Wee Waa is around 150km, which now takes about an hour and a quarter by car on the sealed Pilliga Road. Jim covered the distance on little more than a dirt track with his horse and wagon team, while his wife Ruby followed behind, driving a horse and sulky with five children on board. *Paisley* would eventually support sheep, cattle and a small acreage of wheat, but when Jim purchased it, the land was covered in timber so until it was income-producing he made ends meet as a carrier using his horse team and wagon. From the age of 11, while still attending the small one-building local school, Ray joined his father and older brother Keith working on the property.

'The work was mainly ring-barking the trees and clearing the dead timber,' he wrote. 'Other activities included fencing, lopping trees for stock in times of drought, assisting in the shearing and the dipping and crutching of sheep.'

The family didn't live at *Paisley*. They stayed on a smaller property on the outskirts of town. With Jim and Keith working at *Paisley* from Monday to Saturday morning, it was left to Ray to do most of the chores at the smaller place. He milked the cows, collected firewood, lit the fuel stove first thing in the morning and, in what would become a lifelong passion, tended the garden. He also did odd jobs around the town to earn money, which would lead to him joining the air force.

> In August 1932 with shearing in progress at Paisley, I received a phone call from the local postmaster asking whether I was interested in relieving the post office telegraph messenger who was going on holidays for three weeks. I lost no time in saying yes, and from then until joining the air force I did relieving work in the post office when it was available.

Ray left school at the end of 1931 after gaining his Intermediate Certificate. With the Depression at its worst, there were no fulltime jobs in the town but while working at the post office Ray became proficient at Morse Code and he had dreams of working as a radio operator on an ocean liner. Instead, in September 1938, after seeing an advertisement in the *Sydney Morning Herald,* Ray travelled to Newcastle by train to apply to join the Royal Australian Air Force (RAAF) as a Wireless Operator (Air). His application successful, in April 1939 he was sworn in at the Air Board offices in Melbourne.

Along with around 30 other recruits, Ray began a six-month training course at Laverton air base near Melbourne. His first posting after graduating as an Aircraftman 1st Class (AC1) was with No. 21 Squadron, Melbourne, before joining No. 1 Squadron back at Laverton. On June 30, 1940, No. 1 Squadron, consisting of 12 Lockheed Hudson bombers, took off from Laverton bound for Sembawang Air Base, Singapore. More than 12 months later, the squadron headed to a primitive base at Kota Bharu on the northeast coast of Malaya, close to the Thai border, where they swapped places with their 'sister unit' No. 8 Squadron. The posting was expected to be for six months, but the Japanese had other plans.

'As the months passed the threat of war with the Japanese had increased,' Ray wrote. 'Eventually we were placed on top alert anticipating a Jap landing.'

The squadron carried out reconnaissance flights over the South China Sea looking for a Japanese fleet that was known to have left its base in Indochina. Two fleets were sighted and the US alerted.

The Japanese bombing of the US naval base at Pearl Harbor, Hawaii, at 7.55am on December 7, 1941, was planned as part of a three-pronged surprise attack, with troop landings at Malaya and Thailand scheduled to occur simultaneously. Due to poor weather delaying the air attack on Pearl Harbor, the invasion of Malaya, closely followed by that of Thailand, was actually Japan's first act of aggression of the war. Just after midnight local time, around 5200 Japanese troops began coming ashore on Pantai Sabak beach at Kota Bharu, just three kilometres from the air base where Ray and RAAF No. 1 Squadron were stationed. The Japanese force included three troop carriers supported by a destroyer, the *Ayanami*.

After receiving the go-ahead from Royal Air Force (RAF) headquarters in Singapore, the squadron was quickly in the air under instructions to bomb the ships and strafe the Japanese troops as they came ashore. On the ground, outnumbered Indian soldiers and British artillery attempted to hold back the invasion force with little effect.

The Australians in the air had more success. The first of the Hudsons, piloted by Flight Lieutenant John Lockwood with Bob Thomson, a mate of Ray's from Laverton, as wireless operator, took off in driving rain just after 2am. Ray's aircraft, A16-52, piloted by Flight Lieutenant Oscar 'Ossie' Diamond, a 25-year-old dry cleaner's son from New Farm, Brisbane, with co-pilot Pilot Officer George Evans and rear gunner Sergeant Sam Wells, followed 20 minutes later. Ray, having completed a gunnery course during his time at Kota Bharu, was wireless operator/air gunner. On their first run over the target, they recorded a near miss on a troop ship, but no direct hits. After returning to base to reload with four

250-pound bombs, they managed to take off from the now waterlogged runway and were back at the coast in a matter of minutes. The first ship that came into view was the *Awazisan Maru*, a high-speed cargo ship launched in 1939 by Mitsui OSK Lines for the Yokohama-New York route before being requisitioned by the Japanese Imperial Army as a troop carrier in 1941.

After the war, Flight Lieutenant Diamond, who in 1946 was awarded the Distinguished Flying Cross for his actions that morning, told the ABC what happened next.

> It was the first ship that I saw and straight away we attacked it and did two runs on it. I dropped a couple of bombs on the first run which I think straddled it but on the second run we seemed to have hit it right in the middle with two bombs. There was a terrific explosion, so we knew that we'd done the job on it. At the same time my aircraft got badly damaged from the flak which I suppose came off the bombs because it was a masthead attack. We weren't high level, it was masthead, straight into the shipping.

The official history of Mitsui OSK Lines records the attack from the Japanese point of view and states that the bombs from A16-52 hit the second hatch of the *Awazisan Maru*.

> That ignited the fuel drums and ammunition piled up on the deck into a large fire which spread to the

> bridge and third hatch. All ship function halted but the enemy's attack continued. The order of 'all leave the ship' was issued but there were many wounded crew on the mid-deck. Also, for fear of coming explosions, the landing boats that *Awazisan Maru* had discharged earlier could not approach the ship to rescue. Using lifeboats or jumping in the water, they managed to escape by themselves.

Ray's job at this time was to keep up a barrage of machine-gun fire.

> I had manned the starboard side machine gun and had a busy time spraying the Japs on the deck of the ship and those trying to get ashore.

The *Awazisan Maru* was abandoned at 7am, and its remains finished off by a Dutch submarine two months later. It was the first Japanese ship sunk in World War II and the largest destroyed by Australian forces in the Pacific conflict. Lying in 20 metres of water off Pantai Sabak, it is now a popular diving site.

It was only when they returned to base and ground crew inspected the aircraft that it became evident how close they had come to ending up in the water or jungle.

'Anti-aircraft fire from the protecting Jap cruisers and destroyers was heavy and on our return to base we were not surprised to see we had been hit in a number of places by enemy gunfire, or shrapnel caused by our bombing,'

Ray wrote. 'One missile struck the starboard motor's oil line, and it was estimated the engine would have seized within a couple of minutes.'

Following repairs to their aircraft, the crew was back up in the air five hours later, this time flying some 100km north to Thailand where the second prong of the Japanese invasion force was moving south to link up with their troops advancing north through Malaya. No. 1 Squadron was tasked with bombing a vital railway bridge but caused only light damage. The crew then returned to Kota Bharu where it was the Australians' turn to come under fire, with the base attacked by Zeros and Mitsubishi 'Betty' bombers, and several Australian aircraft damaged by machine-gun fire. With Japanese ground troops advancing to within 250 metres of the airfield, the decision was made to evacuate to Kuantan, 350km south on the east coast of the Malayan Peninsula. When the crew arrived at Kuantan, they found the base in disarray. The aerodrome had been bombed and there were reports of Japanese troops landing on the coast. A friend from the wireless course at Laverton found Ray a bed for the night. He was awakened early the next morning by the sound of gunfire and airmen running and shouting.

> On enquiry I found that the base was being evacuated as the Japanese had come ashore and were not far away. I managed to find a spot in a heavily laden Hudson and we took off for Singapore with the enemy in the bush adjacent to the airstrip. As we rose

> over the airstrip I could see Japs coming onto the runway and some could be seen firing at us.

Back in Sembawang the squadron was billeted in huts among the rubber trees on the perimeter of the aerodrome. Each day Ray's crew flew reconnaissance missions out to sea and up and down the coast. Twice they were attacked by Japanese aircraft, but Flight Lieutenant Diamond took evasive action, and no damage was done. Other crews also came under fire and while no aircraft were lost, one airman was killed and another wounded.

Using the former RAF bases they had captured in Malaya, the Japanese launched regular air attacks on the city of Singapore and it wasn't long before bombs also started falling on Sembawang base. Even so, the squadron continued operations, with Ray's crew part of an attack that caused major damage to a Japanese radio ship, but eventually it became obvious that remaining in Singapore was untenable and the squadron was ordered to evacuate to the island of Sumatra. There were two airfields on Sumatra, known as P1, which was used in peacetime, and P2, a supposedly 'secret' military airstrip 50km west of the city of Palembang where Ray's squadron was sent.

'There were hardly any facilities on P2, it was merely an aircraft landing strip cut out of the bush,' Ray recalled. 'We sheltered in small tents in the scrub, and there was no canteen. I remember that on occasions a beer truck would arrive, and everyone would rush to the truck to get a bottle of warm beer.'

On February 14, 1942, Ray and his crew were involved in a major operation against a Japanese invasion force coming from the direction of Bangka Island off the east coast of Sumatra.

From early morning, RAAF Hudsons and RAF Blenheim bombers had flown missions against the Japanese. Ossie Diamond, with his usual crew of George Evans, Jack Walsh and Ray on board, took off from P2 airfield around 11am.

> We were accompanied by another Hudson until we reached our target. To say that I was amazed at the size of the enemy fleet would be an understatement. The fleet stretched out over many miles and included destroyers, cruisers and an aircraft carrier. Diamond took the aircraft to a shallow dive over a troopship and dropped the bombs, four 250-pound bombs. Some damage was done to the troopship.
>
> After the bomb dropping we looked around for the other Hudson but could not see it. The Captain then decided to go around the fleet again and in doing so was sighted by two enemy Zero fighter planes. The enemy then attacked and we dived to about one hundred feet above the ocean. The attack lasted for several minutes. I manned the port side cabin machine gun, Evans the starboard side gun, Walsh the rear turret and the Captain, the twin Brownings up front. Our plane was badly crippled by enemy fire. The starboard motor was disabled,

> the undercarriage and one wheel being damaged as well as the tail plane.

Reduced to one engine and around 80km off the Sumatran coast and the same again to the strip at P1, and further to P2, Flight Lieutenant Diamond made the decision to try to make it to the closer P1 airfield. The Zeros, believing the stricken Hudson was going to crash, gave up the chase. Diamond followed the Musi River until P1 came into sight.

> Ossie made a terrific crash landing. We had been extremely lucky to survive, especially after he pointed out a large bullet hole near the base of his aircraft seat. He explained that during the attack by the Japs he had frequently risen in his seat to see where the enemy planes were. As luck had it, he was doing this when the bullet struck. We made a hurried exit from the plane until we heard a Pommy voice from a nearby slit hole say, 'Put your f****** heads down as the place is surrounded by Japs.'
>
> Unbeknown to us Japanese paratroops had surrounded the aerodrome that morning. There was plenty of rifle and mortar fire coming from the enemy and British servicemen defenders. Clearly visible were a number of parachutes dangling from nearby rubber trees.

Diamond decided to try to fly out another damaged Hudson located near the hangars. The crew ran over and clambered aboard, but the tips of the propellers had been bent in a forced landing earlier in the day. Despite Diamond gunning the engines, the aircraft would not lift off and came to rest against barbed wire at the end of the runway. They set it alight using a flare gun and made a run for the rubber trees, about 100 metres away.

> Before reaching the rubber trees we were subject to rifle and mortar fire from the Jap paratroops but fortunately not one of us was injured. Tall grass grew among the trees and there we sought cover. Crouching low in the grass we could hear a group of Japanese talking a short distance away but our luck held and we were not spotted. When dusk fell, we joined a group of British soldiers and headed for Palembang via an area of swamps and paddy fields. Leeches clung to our legs, and I lost my shoes in the bog. Sometime after midnight we reached the city and headed for the Air Force headquarters. The officer on duty gave me his size nine tan shoes to wear and which I kept for about two weeks. My normal shoe size is six and a half.

The next morning Ray and Jack Walsh caught a bus to P2, only to find the squadron had left for Java the previous day. Ray went to his tent, but all his clothing and some money he had left with a friend for safekeeping were gone. He and Walsh caught another bus to Oosterhaven, a port on the eastern end of Sumatra. They then caught a ride on a boat to Batavia (now Jakarta) and obtained military transport to RAF headquarters at Semplak, near Buitenzorg in central Java.

Told that what was left of No. 1 Squadron was now based at Kalijati on the north coast, near to where Japanese forces were reported to be planning a landing, Ray caught another bus to the town. He arrived at the airstrip after nightfall to find that his squadron was engaged in trying to bomb the enemy a little distance offshore. He was told to report for flying duties the next morning and found a bunk for the night.

> I was awakened early in the morning to hear that the Japs had landed and some had been seen near the perimeter of the airstrip. The Commanding Officer then gave the order to abandon the station. Along with others I hurried to the aircraft hangars and went aboard a Hudson about to take off. As we became airborne we were fired upon by Japanese troops in armoured cars near the runway. Our destination was Bandung in central Java.

When they arrived at Bandung and joined up with No. 8 Squadron, the hopelessness of the situation became obvious. Combining their resources, the two squadrons could muster just a handful of aircraft. It was decided that all No. 8 Squadron personnel and non-vital staff from No. 1 Squadron should be repatriated to Australia, leaving around 180 men, mainly aircrew and ground engineers.

The remnants of No. 1 Squadron did their best to remain operational and continued to fly reconnaissance missions despite overwhelming Japanese opposition, but when they were down to just three serviceable aircraft the squadron's commanding officer, Wing Commander Reginald Davis, decided it was time to leave Java.

> He directed that we travel by trucks to a beach on the south coast near Tjilajap where Catalina aircraft from Australia would pick us up. As we waited in our trucks near Bandung, a Hudson flew low over us heading for home. We were all very miserable.

Ray and about 30 others stayed in an abandoned hut on a tea plantation a short distance away in the hills. The remainder of the squadron was camped on the beach. One of the wireless operators had set up a transmitter and got through to the RAAF in Darwin, who advised that Catalinas would be sent from Broome, Western Australia, to pick them up. When the aircraft hadn't arrived after two days, Ray was part of a party sent further east along the coast to check if they had landed there.

> After many hours traversing the rough country we reached the ocean at Tjilajap and waited. We stayed there a couple of days and no Catalinas arrived so we returned to the tea plantation. Several more days passed and it became obvious that we were not to be rescued.

What Ray would not learn until after the war, was that on March 3, 1942 — the very day that he and the others were scanning the skies hoping to see the flying boats heading towards them — Japanese aircraft had attacked Broome Harbour, destroying six Catalinas and killing 80 Dutch civilians, including women and children, who had just been evacuated from Java.

'The next thing to occur was the receipt of a message from the Japs that we were to give ourselves up or they would come and take us,' Ray wrote. 'Faced with the futility of trying some way of escaping, which had been given considerable thought, we boarded our truck on 6th March, 1942, and came to a small town called Wanaraja. There we passed a line of Jap troops and became POWs. I had a very sinking feeling. I think the official date of my capture was 8th March, 1942.'

AFTER A MONTH IN camps at Wanaraja and the town of Leles, Ray and his fellow POWs — including two close friends from his squadron, fellow wireless operators/air gunners Dick 'Rocket' Head and Keith 'Killer' Guthrie — were moved by train to Jakarta. From the station they were marched to a former Dutch army base known by the prisoners as 'Bicycle Camp' because of the number of bike racks lined up outside the buildings. Living conditions were quite acceptable, but their treatment by the guards gave the POWs a small taste of what was to come.

> There were some hundreds of Australian POWs incarcerated in the camp, mainly those soldiers from the 7th Army Division, and others from the 8th Army Division, who had escaped from Singapore. Sailors from the Australian warship HMAS *Perth* were also there. The Aussie warship had been sunk earlier in a sea battle with the Japanese Fleet in the Sunda Strait. We all had to sign a document declaring we would not attempt to escape, which we signed under duress.
>
> Rice and vegetables were in reasonable supply but from time to time there were uninvited supplements in the rice. The Japanese soon had us out working. Most of our time was spent rolling petrol and kerosene drums around the wharf at Tanjung Priok, the port for Jakarta, some distance from the city. We were also required to dig air raid trenches in the parks around the city. The Japanese at all times were hostile towards us and because we had become POWs regarded us as poor specimens. It was not in their code to become prisoners and indeed preferred death to that. There were many bashings with rifle butts but nothing like what was to happen along the Burma Railway.

On October 11, 1942, seven months after surrendering to the Japanese, Ray, Rocket and Killer were among 1200 Australian, Dutch and American POWs taken by open truck to Tanjung Priok, where they were shoved into the holds of a rusty old Japanese freighter, the *Dainichi Maru,* bound for Singapore.

> It was most oppressive from the heat, body against body, and barely enough room to lie down. The food rations were very scanty and consisted of a small helping of rice with a little watery vegetable soup. The toilets were up on deck and consisted of rails slung over the side of the ship. The equator line runs between Indonesia and Singapore and the humidity was almost unbearable, and we became wet through from our own perspiration. The Japs were ever present, shouting and screaming at us in Japanese and broken English. A thump with a rifle butt was very common. The boat journey took about three days.

On arrival in Singapore, they were taken by truck to the former British Army base at Changi that was now a series of POW camps run by Japanese guards supported by Indian troops who had defected after the fall of the island. Ray and his group were housed in a badly damaged cottage, with rice and vegetables heavily rationed. Each day the prisoners were put to work loading and unloading ships in Keppel Harbour.

Every morning and afternoon, as they headed to and from the dock, they would pass a high-rise prison that had been taken over by the Japanese solely to house British, Australian and Dutch civilian detainees. Often Ray and his fellow POWs would see their hands coming through the bars, sadly waving at them. It all added to the misery of their situation, but there was at least one memorable event that broke the drudgery and gave the Aussies something to laugh about.

> One day at the docks a German submarine was in port. A German sailor came off the submarine and became engaged in conversation with a couple of our fellows. The Japanese guard started screaming at the German, obviously thinking he was another POW. The German, without a word spoken, took hold of the Jap and pushed him off the wharf into the water below.

It was a rare light-hearted moment for the prisoners as the weeks turned into months. As their first Christmas in captivity rolled around, Ray and some of his fellow airmen were invited to 'celebrate' in the quarters of an 8th Division army colonel.

'It was hardly a jolly affair,' he remembered.

On January 9, 1943, Ray was on the move again. Gathering up his few belongings — consisting of two pairs of shorts, two shirts, one pair of Indian Army boots obtained in Java, a metal 'Dixie' eating dish, spoon, water bottle and a piece of canvas to wrap it all in — he joined with a large contingent of prisoners who were taken to Singapore railway station and transported by train to Penang where they were crammed into the hold of another old Japanese freighter.

> About two days out from our destination our convoy of freighters, including one filled with Jap soldiers,

> was attacked by US Liberator bombers. The ship with the Jap troops was sunk and our ship escaped with a near miss. The miss must have been very close as red-hot pieces of metal whistled just over our heads when in a prone position in the hold. Several of the POWs were hit by shrapnel, including a chap named McCredie, who was fatally wounded. He was connected to a Wee Waa family of the same name. The Japs, who were up on deck manning machine guns, were all killed. The Japanese soldiers from the sunken ship were taken aboard our vessel.

They arrived at Moulmein in Burma (now Myanmar) on January 17, having made their way up the Salween River. During a brief period on deck, Ray saw the corpses of numerous Burmese civilians floating towards the sea. The POWs were marched through the city to an old jail. Looking through the bars of their cells they could see the magnificent golden Kyaikthanlan Pagoda, immortalised by Rudyard Kipling in his poem *Mandalay*.

> Ship me somewhere east of Suez, where the best is like the worst,
> Where there aren't no Ten Commandments an' a man can raise a thirst;
> For the temple-bells are callin', and it's there that I would be —
> By the old Moulmein Pagoda, looking lazy at the sea.

There would be no looking lazy at the sea for Ray and his fellow POWs, just cruelty, deprivation and, in many cases, death. They were about to become forced labour on one of the most infamous undertakings in modern history: the Burma-Thailand Railway.

THE US NAVAL VICTORY in the Battle of Midway in early June 1942 greatly reduced Japan's ability to supply and reinforce its land forces in Burma by sea. As an alternative supply route, the Japanese decided to build a 415km railway between Burma and Thailand. All but 50km of the line needed to be cut through dense jungle and required construction of more than 600 bridges and hundreds of viaducts. To carry out the work in a near-impossible 12-month timeframe, the Japanese relied on the forced labour — known to them as 'romusha' — of about 200,000 local men, women and children, and 60,000 Allied POWs, of whom 13,000 were Australian.

For Ray and his fellow POWs the nightmare began at the village of Thanbyuzayat, where they were left in no doubt of what was expected.

'We were addressed by a senior Japanese officer who told us that the line would be completed, no matter the human cost, and go over our bones if need be,' he wrote.

Rather than start from both ends of the route and meet in the middle, the Japanese chose to set up hundreds of camps at various points along the way, with each camp responsible for its section of track before moving up the line. Ray's group was taken by bus to 18 Kilometre

Camp, where they joined up with a diverse collection of prisoners. As well as Ray and his fellow airmen from No. 1 Squadron, there were Australian infantrymen from 7th and 8th Divisions, sailors from HMAS *Perth* and USS *Houston*, soldiers from a Texas artillery unit captured in Java, and Dutch Army personnel who were largely Javanese. They were designated Prisoner Group No. 5, and later No. 3.

The camp, typical of most set up along the route, comprised several bamboo huts with thatched roofs, each about 20 metres long. There was a central aisle running from end to end and on each side were raised bamboo platforms about 60cm off the ground on which the prisoners slept. There were no washing facilities and the toilets consisted of open slit trenches. The kitchen was manned by prisoners. At first, the rations were acceptable, consisting of a cup of rice with a small portion of vegetables, or a weak vegetable or meat-flavoured stew.

The first morning in camp, Ray and his fellow POWs were introduced to a daily routine they would get to know all too well.

> At first light we were fed and then marched out along the line for some distance to commence our day's toil. Accompanying us were Korean guards and a number of Japanese engineers. The Koreans were real thugs and bashings became common place. To retire to the bush with a shovel for toilet purposes, one first had to

> salute the Korean guard and say 'Benjo Nippon', and on return, again salute. Our task on this section of the line was to excavate clearings and build embankments where necessary.

At first the prisoners were required to remove one cubic metre per man per day. Some did the digging, others shovelled it onto bamboo matting attached to poles, and two more prisoners carried it away. It was hard work and soon the men were covered in dust and soaked in sweat. The Australian officer in charge of each group warned the men to make their one water bottle last for the day, because they wouldn't be getting any more. Early on the POWs set themselves a target of completing their one cubic metre per man target quickly so they could get back to camp early in the afternoon, but the 'early mark' didn't last long.

> The Japs soon fixed this by increasing the workload from one cubic metre to two cubic metres per man and then to three cubic metres and so on. This meant we were leaving camp in the early morning and returning near dark. On the way back to camp there was a muddy water hole and each afternoon we had to wait while the working elephants first had their bath and do what elephants always do. We were then permitted to bathe in the filthy water.

The prisoners' next base was 80 Kilometre Camp, but in May they were moved on foot another five kilometres up the track to a camp that proved unworkable due to its swampy conditions. They would only stay a few weeks before being marched back to Camp 80.

> By now many POWs came down with malaria. I was one of the first to get it and was very ill. Throughout my stay in Burma, I frequently had this illness. Others had horrific ulcers on their legs, and dysentery. There was practically no medication to treat these diseases.

The Indian army boots that Ray had managed to obtain in Java were to prove a life saver.

> They helped me to avoid the sharp bamboo stumps and other hidden dangers on the jungle floor and so minimize the chance of getting a tropical ulcer. These ulcers grew rapidly, and you were lucky to survive after having a leg amputated. The amputation was carried out by an Australian doctor without anesthetic and most patients died from loss of blood or shock.

By the time the prisoners were moved further into the jungle to 100 Kilometre Camp the wet season had started, and the working conditions became even harder as the number of prisoners cut down by disease increased. Rations were allocated only according to the number of railway workers. Ray recalled that the Japanese had a strict criterion for assessing whether a prisoner was fit for work.

'If you could stand up, you were fit,' he wrote.

There was no formed road near the rail line, only a rough track through the jungle. With the rain tumbling down and the workers dressed only in shorts and boots, the task of building a track for the railway became near impossible.

> The number really fit for work had plummeted and many bashings took place against officers trying to protect the sick men. The officers were belted with rifle butts if the Jap engineers considered more of the sick were able to work. The ulcers on some of the prisoners' legs were huge and grew very quickly. Without medication there was little that could be done for them. After a while the doctor in our camp, an American from the Texas artillery, Dr Lumkin, started to scrape away the putrid flesh each day with a small spoon causing extreme pain for the patient.
>
> The stench from the sick hut was shocking. A number of POWs were dying daily, including our doctor, and the last post seemed to play incessantly. Cholera also struck the camp, and the Japs became very panicky, spreading lime everywhere. Dysentery was rife in the camps and with ulcers, pellagra, beriberi and malaria, most of the POWs became living skeletons. I do not remember the number of times I went down with malaria and beriberi. With malaria comes a powerful fever and violent headaches. It took Herculean-like efforts just to push a little rice down. It rained practically non-stop for weeks. The huts

> leaked but, despite all this, the Japanese insisted on having a quota of men to work on the line each day.

Those considered too sick and frail to work were either left to die or sent by truck to Thanbyuzayat where, according to the Japanese, they would be treated in hospital. Few were ever seen again, and when Ray's great mate Rocket was loaded onto a truck along with other desperately sick prisoners, both men feared the worst. Somehow, Rocket defied the odds and returned a few months later.

Throughout it all, there was one constant — the barbarity of the guards.

> The Japs were cruel and sadistic beyond belief. One day a group of Japs arrived at our camp with two British prisoners roped together. They were left for hours strung up by their arms on a line above the ground. The Japs were parading the two escapees as a warning about trying to escape. The prisoners had managed to get some distance from their camp on their way to India but were betrayed by local people. They were later executed in Thanbyuzayat.
>
> The Korean guards were a vicious lot and seemed to get pleasure in bashing POWs without any just cause. They had learnt well from their masters.

According to the Australian War Museum, POWs on the Burma-Thailand Railway would give their guards 'inventive' nicknames.

'Nicknames remained one of the few ways prisoners could retaliate against the men who controlled their lives,' the museum notes in its exhibition, *Stolen Years — Australian Prisoners of War.* Nicknames listed by the museum include The Boy Bastard, The Boy Bastard's Cobber, Fishface, Poxy Paws, Babe Ruth, Gold Tooth, Paddle Feet, The Snake, Modern Girl, Maggot, Boofhead, Snake Eyes, Charlie Chaplin, Barrel Guts, Wire Whiskers, Tom Mix, and Foghorn.

Ray found from painful experience that one of the nicknames was particularly well earned.

> In Burma, the Korean guard nicknamed the Storm Trooper, who was tall and dark, took a dislike to me and three or four other POWs as we lined up at first light to go out to the railway line from the 100 Kilo Camp. He was a well built man with a punch like that from a heavyweight boxer. Down I would go with a bloodied nose or painful mouth. I soon learnt not to get up straight away as I would be quickly knocked down again.

The guards would occasionally dish out light punishment to their own lower ranks in the form of a slap or two to the face. One day, Ray witnessed an example of Japanese army discipline when working outside the 100 Kilometre Camp during the wet season.

> The Jap army were moving field artillery towards the war front in Burma. The artillery guns were

> being hauled by teams of six to eight Jap soldiers in harness. It was a very rough track on which they were moving and the rain was pelting down. The team I am about to discuss was only about 30 to 40 metres away from where I was working. One Jap in the harnessed team collapsed in the mud. The Jap officer, all neatly attired in his uniform with a sword at his side, came up to the fallen soldier and issued commands in a loud voice. There was no response from the soldier, and to my surprise the officer drew his sword and belted the stricken soldier with the blunt edge of the sword. The soldier did not move during this punishment and the officer then gave the order to unhitch the soldier and to proceed on their way. Some 20 minutes or so later the soldier rose from the mud, shook himself and trotted away to catch up with his team.

Towards the end of 1943 the rail line was nearing completion and the Japanese demanded the work rate increase, despite the deteriorating condition of prisoners who would be forced to provide manpower for a wide variety of forced labour. On one occasion, Ray was part of a detail ordered to push a broken-down truck for over a kilometre along a rough jungle path to a depot. Other times, when constant rain made it impossible for trains to climb a hill near their camp, up to 50 prisoners would be sent out in the middle of the night to push the locomotive and carriages up the steep gradient.

With the end in sight, work on the railway intensified, often late into the night, making conditions even more perilous.

> I might say it was very scary working at night in poor light on a wooden bridge twenty to thirty feet above raging river streams. At other times we were required to stand in the river pulling ropes to the count of ichi, ni, san, shi (1, 2, 3, 4) and then let the rope go and so allow a wooden log to drop onto a pylon in the river.

Just when the prisoners felt they could give no more, they were forced to increase their output.

'We were leaving the camp at daylight and not returning until ten or eleven in the evening,' Ray wrote.

It was a time of incredible physical and emotional stress and pain, illness and cruelty. Even so, through everything they endured at the hands of their captors, Ray and his fellow prisoners managed to hold on to something that could never be taken away from them: their spirit.

> I shall always remember the ditty, 'She'll be coming round the mountain when she comes', which we sang marching back to camp at night.

THE BURMA-THAILAND RAILWAY was completed in October 1943, but the Japanese still had big plans for their POW workforce. After two months in a camp close to Three Pagodas Pass near the Thai border, the sick and dying in Ray's group were moved to Thailand. The remainder were combined with another group and sent by train along the track they had just built to a camp across the border at Tamarkan, close to the bridge on the River Kwai. On the way they passed over the notorious 'Pack of Cards' trestle bridge that had collapsed three times during construction.

'We thought ourselves fortunate not to have worked on it,' Ray wrote.

At Tamarkan, Ray, Rocket and Killer were among 600 prisoners chosen to be given a rudimentary medical examination and asked for their occupation prior to the war. As first of the three in line, Ray answered, 'Farmer.' Wanting to stay together, his two mates gave the same answer. On March 27, 1944, they were loaded onto a train, about 30 men crammed into each small goods wagon.

It was the start of an exodus on rail, boat and truck that would last six months and include stops in Bangkok, Cambodia, Vietnam and Singapore before finally ending in Japan. In Saigon (now Ho Chi Minh City, the largest city in Vietnam), the POWs were held in a former army base where rice and vegetables were in good supply and the Japanese commandant, who spoke reasonable English, claimed to have attended Sydney University. Not that his links to Australia softened the attitude of the guards in charge of the daily work details. One day, Ray was put in charge of a group digging slit trenches at the airfield. When the guard felt the workers were moving too slowly and his cries of 'Speedo, Speedo,' were being ignored, he took it out on Ray, who was given a beating with a rifle butt and a stick.

On their return to Singapore the prisoners encountered new horrors.

> We arrived at Singapore Rail Station at night on 4th July 1944, and moved to River Valley POW camp in the city. It was a large camp with Australians, Americans and British already there. Most of the train travellers marched to River Valley but because my feet were in bad shape with sores, blisters, and with severe beriberi, I went by open truck. Getting to the camp early allowed me to select three good spots in one of the huts for Rocket, Killer and I. The camp was a filthy and run-down place. It had been used as a main transit camp with an abundance of rats and scrawny cats.

Ray witnessed starving prisoners trying to catch the cats on occasion, presumably for food, but never reached that desperation.

> Hungry and starved as I might be I did not fancy a cat meal. However a near neighbour caught a large snake and I participated in the eating thereof. I did find it was quite tasty, a bit like fish.

For weeks the prisoners were transported across Singapore's Keppel Harbour by barge to and from a small island they called 'Jeep Island', where they worked day and night excavating a large area for construction of a ship repair base. This work ended early in September when the POWs were assembled and informed by their guards that, 'All men go to Nippon.'

In the days prior to the announcement, as the prisoners had crossed Keppel Harbour, they had spotted the *Rakuyo Maru* lying at anchor and correctly surmised that this was the ship that would transport them to Japan. Ray, Rocket and Killer were told that they would be boarding the ship ready for departure in a convoy on September 6, and they gathered their few belongings together for the journey. At the last moment there was a change of plan.

> On the evening of 5th September 1944 the Japanese told us that the last three sections of Australians numbering 40, 41 and 42 would not go and instead they were taking Englishmen from the other side of the camp. I and my two mates were in Kumi Section 40. The three sections left behind numbered roughly one hundred and fifty men.

On September 12, 1944, the convoy was attacked by three US submarines in the Luzon Strait between Taiwan and the Philippines. *Rakuyo Maru,* with 1300 POWs onboard, was sunk by USS *Sealion ll*, and the *Kachindo Maru*, which contained 900 British prisoners, by USS *Pampanito*. A total of 1159 POWS died, some 350 of them when the lifeboats in which they were rowing towards land were attacked by a Japanese navy vessel the next day. A few other lifeboats were picked up by a Japanese ship, and the three US submarines that had attacked the convoy later returned to the area and rescued another 149 survivors who had been clinging to rafts and wreckage for three days. Of the 717 Australians on the *Rakuyo Maru*, only 174 survived. Among those who lost their lives were Brigadier Arthur Varley, the officer commanding the 9000 Australians working on the Burma-Thailand Railway, and rugby footballer Winston 'Blow' Ide, who played two Tests for the Wallabies against the All Blacks in 1938.

For Ray and the 150 or so other prisoners who were held back in Singapore and so missed the carnage of the *Rakuyo Maru*, life continued as before, loading and unloading Japanese cargo ships on the docks of Keppel Harbour. After three months another ship, the *Awa Maru*, was ready in dock to take the remaining prisoners to Japan.

> On 24th December 1944 we were marched to the docks area and boarded the *Awa Maru.* We were again herded below deck with body-to-body contact as we huddled on the floor of the ship's hold. The ship was packed tight with several hundred British and Australian POWs plus cargo, mainly rubber. The heat was most oppressive, and the toilet facilities were again non existent or very primitive. A timber railing was slung alongside the ship and to use the toilet we

> sat on the rail. On Christmas Day we stayed put in the dock. There was no joy that day I can assure you.

The *Awa Maru* was part of a large convoy that included a light aircraft carrier. Leaving on Boxing Day, 1944, it arrived at the northern port of Moji on Kyushu Island on January 15, 1945, after narrowly avoiding a torpedo attack off the coast of Vietnam. For the prisoners, who were now used to being human workhorses for the Japanese, the only difference in their situation was the climate. As the *Awa Maru* moved closer to Japan the POWs sensed a gradual drop in temperature. A few days out from Moji it began to snow. To the prisoners' astonishment, the guards handed out brand new greatcoats, plundered from the British Army stores in Singapore.

After being offloaded, the POWs were marched through the main street of Moji, watched silently by locals who lined both sides of the roadway. After standing for hours in an open square they were moved into a large hall. At around 7pm, having not eaten all day, the starving prisoners were each given a bread roll and told not to eat it straight away as they were soon to embark on a long train ride. The bread rolls were all downed immediately. After several hours travelling south on the island of Kyushu, they arrived at Omuta, about 60km northeast of Nagasaki, where they were taken to their camp by bus. No one had told the prisoners what work they would be doing in Japan, but it didn't take them long to work it out. As they arrived at the camp, they saw American prisoners leaving what appeared to be a mess hall, all with blackened faces. The newcomers began whispering among themselves … they were about to become coalminers. Ray described his new home.

> The POW camp was known as No. 17, Fukuoka Prefecture, and one of the largest in Japan with more than 1000 mixed nationality POWs. There were numerous timber huts each holding about 40–50 men. Each hut was divided into five or six rooms having about ten men in each. A passageway ran down inside of each building with water taps and toilet pits at the ends. Running along the side of the passageway was a raised platform on which one had to line up to be counted every morning and evening, numbering off ichi, ni, san, shi and so on. One prisoner could not count in Japanese and the guard used to rant and rave in Japanese and sometimes jabbed him with his rifle.
>
> Big counts of all prisoners took place frequently in an open square, with counts and recounts on an abacus. The huts had wooden floors and the straw bedding on the floors was quite comfortable and the bed covers kept us warm. The camp was situated near the sea with a high paling fence topped with electrified wire.

The men were divided into work groups of about 50 each. Every morning, they would be fed a level cup of rice, garnished with a sprinkle of seaweed and a few pickled vegetables. The same would be provided in a small wooden container known as a 'binto box' for lunch down the coal mine. The mine was about 750 metres from the camp and comprised working levels at 1200 metres and 1800 metres under the seabed. The POWs worked at 1200 metres in around-the-clock shifts. They were taken down to the work-level in a train driven by a conveyor belt and issued with miners' helmets with lamps.

> My POW number was 1583 and each day I had to yell out this number in Japanese when receiving or returning the lamp. Every day was spent either shovelling coal into skips, repairing ceilings and train lines, or proceeding individually down low tunnels, about a metre and a half high, to collect timber poles from a railway point in the mine. The task was very arduous to say the least. Each pole was two and half to three metres long and 12 centimetres thick and three or four tied together had to be hauled along the mine floor. The haulage distance was 50 to 60 metres.

From time to time the Japanese engineers would move around testing the ceilings and walls — and tormenting Ray.

> They were armed with steel rods about 1.5m long and took great delight in spearing me around the rib cage. At that time I was very thin and my ribs were well exposed. It was extremely painful and they thought it a great joke.

Dressed only in g-strings below ground, the men would be covered in black grime from head to toe and with no way to bathe properly back at the camp, Ray would stand underneath the leaking ceilings at the end of his shift in an attempt to wash off some of the dirt.

> Personal hygiene was a real problem. Ash was used to try and clean teeth. Toilet facilities were non existent in the mine. At times this was most uncomfortable when in the mine for 8 to 12 hour shifts. No soap or toilet paper was provided. We must have been very smelly. However, throughout our period of captivity our head hair was kept clipped right off and although all razors were taken from us they allowed a couple of POWs to run a barber's 'shop'. Mind you, the shaves were pretty rough as the razors dragged a bit. No beards or long hair were allowed by our captors. We were not bothered by lice or bed bugs, unlike previous camps in Burma, Thailand and Indo-China, now South Vietnam. In Burma I spent part of my lunch break debugging the lice from my clothing and having a good scratch of my head.

Ray's health worsened in Japan with malaria, beriberi and bronchitis recurring regularly. He was emaciated and felt himself getting weaker by the day. For the first time during his captivity, he began to wonder whether he would survive, although his determination to make it back to his family never dimmed. He went to the

Australian army doctor, who had the task of tending to the POWs and was given a few days off work. Finding it difficult to force down his rations of dry rice and watery soup, his health continued to deteriorate to the point that he received a visit from Bob Thomson, his friend from Laverton who had been on the first aircraft to attack the Japanese in Malaya.

> He gave me a pep talk and a warning to push the rice down no matter what. Somehow, I survived this extremely tough period. So much so that the Japs insisted I had to go back to the mine.

He was far from well. Racked by severe beriberi, one night after a long shift down the mine he struggled to walk back to the surface and then had to be helped back to camp by his mate Killer. He found it hard to sleep due to bronchitis, which made it difficult for his fellow POWs as well.

Ray's nemesis, the sadistic Korean guard known as the Storm Trooper, had accompanied the POWs on the *Awa Maru* to Japan. While Ray had managed to avoid him on the voyage, the Korean had made his presence painfully felt by other prisoners. Fortunately for Ray, he did not see the Storm Trooper again after the ship docked in Japan, but that did not mean he was safe from the cruelty of other guards.

It was in Omuta that he received his worst beating of the war.

> I was coming back from seeing the doctor, who incidentally had no drugs to help, and crossing an open area with the guardhouse about 50 metres away. With me was a POW roommate called Scotty. As we drew to eye level with the six guards sitting on the front veranda of the guardhouse, I realized it was time to make our salute to the enemy. I mentioned this to Scotty and he said not to bother about it as we were so far away. Foolishly accepting that advice, I did not salute but had only gone another pace or two when a loud voice from a guard rang out 'kootchacoy' (come here). The time was about 1pm and it was around 6pm when we were allowed to leave. We were in a very sick and sorry state when we left the guardhouse.
>
> Firstly they belted us across the backside with a piece of timber resembling a baseball bat. We had to come to attention each time we were knocked down. After being felled five or six times the guards then gave us the choice of more of the baseball bat or fists. It was a very severe beating and the guards were delighted. My face was bloodied and bruised and likewise my back and backside. I was very sore for days afterwards.

The weather conditions made living and working in Omuta even harder as the men toiled through a harsh winter. For weeks, the prisoners never saw sunshine, just the occasional glimpse of a red ball in the murky sky. That began to change in March 1945, with the first signs of spring, and it was not just the milder weather that caused the POWs' spirits to rise. In May, air raids by US bombers in the area became more regular, indicating that Allied forces were operating closer to Japan and that the Japanese defences were becoming less effective. Many nights were spent in cold underground air-raid shelters as US incendiary bombs rained down on the camp. A number of POWs were killed, and part of the camp's electrified fence was destroyed.

When the raids continuing unabated through July and into August with little resistance from the Japanese, the POWs felt the war must be coming to an end. But it gave them little joy.

'We assumed that when the Americans did close in we would be shot or bayoneted,' Ray wrote. 'On the way to the mine we often saw 12 to 14-year-old boys practising with wooden rifles.'

One night early in August, Ray returned to the camp after a day working underground to be told that the sound of a large explosion had been heard coming from the direction of Nagasaki. Six days later, on August 15, 1945, he and his work team were told to down tools and were taken to the surface several hours earlier than usual.

> On arrival at topside we found other working groups of POWs had arrived there earlier than usual. When all POWs reached the top we were marched back to camp about 3 o'clock in the afternoon. On the way we heard that Red Cross parcels had surfaced and we

> were to share a parcel between six men. That night I 'feasted' on Spam and very mouldy cheese. They were delightful.

THE JAPANESE NEVER TOLD the POWs that the war had ended. They were left to figure it out for themselves. First, the Allied bombing stopped. Then the Japanese guards disappeared. It wasn't until about 10 days later, when a representative of the Red Cross arrived at the camp that the prisoners were given the news. When he told them about the atom bombing of Hiroshima and Nagasaki, the significance of the loud noise that the men had heard two weeks earlier became evident.

'It is extremely difficult to express my feelings at that time,' Ray wrote. 'Certainly it was one of high elation and with a sense of good luck and thankful of my survival. My health had not improved but I had made it. My mind constantly went back to my parents at Wee Waa and the hope they were alive and well.'

The war might have been over, but it would be a month before the prisoners could begin the journey home. During that period some of the POWs who were well enough ventured outside the gates to look around or gather clams and mussels on the beach. US planes made regular food drops, with Spam, cheese and chocolates in plentiful supply. American troops also tracked down some of the Japanese guards and mine supervisors who had made the prisoners' lives hell and brought them back to the camp, where they were put to work at humiliating tasks such as cleaning out the latrine pits.

'Occasionally one would see a POW flatten a Jap,' Ray recalled. 'Not for me. I just wanted to get home.'

Finally, the day came. A large number of prisoners, including Ray, Rocket and Killer, were taken by bus to Omuta railway station and then by train to Nagasaki. They were overjoyed to be leaving the camp for the last time. Especially as an interpreter had informed them that the Japanese originally planned to kill all prisoners of war after the first atom bomb had landed on Hiroshima. They had been saved only by Japan's decision to surrender immediately after the bombing of Nagasaki three days later.

The sight that met the prisoners as they stepped from the train at Nagasaki was one that Ray would never forget.

> I was completely shocked at what I saw of Nagasaki. Excluding the hill area and the waterfront, the city had been flattened. All one could see was twisted concrete and metal.

Waiting for them at the waterfront was the USS *Cape Gloucester*, a newly commissioned aircraft carrier that had been in service only three months before being ordered to join what became known as the 'Magic Carpet Fleet' that returned POWs from Australia, New Zealand, Britain and Holland to their homelands.

Before they boarded the ship, the men were served donuts and coffee by WAACs (members of the US Women's Army Auxiliary Corps). They then went inside the wharf building where they discarded their clothing, showered and were disinfected before being issued US Army uniforms.

The *Cape Gloucester* took the prisoners to Manila in the Philippines, where they stayed for six days before being flown by RAAF Catalinas to Moratai, Indonesia, and then, finally, to Darwin, Australia.

'It was a magnificent feeling to reach that city,' Ray recalled. 'Almost unbelievable to me.'

The men stayed in Darwin for three days. Ray called his mother from his hotel. The line was so bad that they conversed with the help of the operator at the Wee Waa telephone exchange who acted as an intermediary.

Ray said of his time in Darwin that he would always remember the smell of the gum trees. 'It made me realise that I was truly back home in Australia,' he wrote.

With transport arranged to take the men back to their homes around Australia and New Zealand, it came time for the three mates, McMillan, Guthrie and Head, to go their separate ways. Guthrie stayed in Darwin for medical tests on his lungs, Head took off for Adelaide and Ray boarded a RAAF Liberator bound for Sydney. The plane landed at Cloncurry in Queensland, where the locals had prepared a welcoming feast at the Town Hall while the plane was refuelled before the journey continued.

And so at last we reached Sydney. My worldly possessions at that time were the American Service uniform I was wearing and a carton of Lucky Strike cigarettes.

I had wondered who would meet me at Mascot. I was directed to a small wooden building and waiting for me were my sisters, Nell and Evelyn. What a joyous occasion, and I will remember the meeting for many a day to come. My sisters told me our mother was waiting for me in a hotel near Central Railway. On meeting my mother, we were both overcome with emotion. Tears flowed freely from each of us and we embraced one another for a long while. We were both so happy. On going downstairs to buy a drink I received a great welcome. It was an occasion when I did not have to hit the pocket nerve to pay for the drinks. I spent the arrival night in the hotel and next day my mother and I journeyed by train to Muswellbrook where my sister, Grace, and her family lived. I was given a great welcome.

After two days my mother and I continued on our way by train to our home town. Nearing the town, I viewed the old home which is not far from the railway line. It seemed just as I had left it more than five years before. My father was at the railway station to meet us. He seemed to me to have changed little. On the Saturday afternoon he asked me to join him in a beer at the Imperial Hotel. I had never been in a pub with him before. He told me that they never expected me to return. The only news that they had of me were the official notification that I was missing, believed to be a prisoner of war, and a small post card sent by me from Burma when I was working on the railway line. The card is still in my possession. One could not say in the card what the real position was in regard to health, provisions and treatment, otherwise news of me would not have reached home.

Some little time after arrival back home, the locals gave me an official welcome home in the School of Arts. In all I had been away overseas nearly five and a half years. Such a length of time passes quickly in Australia but overseas, particularly as a POW, each month felt like one whole year. I was determined to see it through and believed I would make it. When I left Australia in 1940 I weighed 10 stone and 10 pounds, or 68 kilograms. When I returned in 1945 I was less than half that weight. For some weeks while at home I grew a partial beard because I was so thin and the razor kept cutting into my cheek bones. My life as a prisoner of war was a day-to-day proposition with mateship and help from close companions in the battle to survive.

The advice given to all POWs returning home was to forget about our experiences and carry on living as before. No counselling of former prisoners was given and some could not adjust to civilian life. A friend of mine took his own life shortly after returning. I spoke little of my imprisonment for many years and it is only

> in recent times that I can talk about those years with the Japanese. I can become quite emotional at times.
>
> The harshness, brutality and lack of compassion by the Japanese are indelibly imprinted on my mind.

And with those words, the man next door ended his memoir. However, there is a postscript.

In 2016, my 15-year-old son Will and I took a road trip from our home in Queensland to Sydney. We visited friends, took in the tourist attractions and visited some of the haunts where my wife and I had spent our time before heading north for a new life. Our last stop was the neighbourhood where I had grown up. A lot of the old houses were gone, having been knocked down and replaced by large, modern edifices. Our little home was still there, looking much the same as when we lived in it, as was the Suttons' and the Donalds'.

The McMillans' house was untouched as well, still white and bright and gleaming in the sunshine. There was one thing I couldn't help noticing though.

The new owners — an Asian-Australian family — were out the front, washing their car. Parents and children, five or six of them in total.

It was a Japanese car, a Mitsubishi Outlander SUV, and it was parked not on the driveway, but sideways across Ray's beautifully manicured lawn. I'm not sure if I just thought it as I slowly cruised by, or said it out loud, but the words are just as clear in my mind now as they were at that moment.

'What would Mr McMillan think about that?'

I was to get my answer a few years later.

When I sent a first draft of this article to Ray's son Stephen, he voiced his concerns that the final paragraphs might give the impression that his father was racist in his later years.

'He was not,' Stephen wrote. 'Indeed, you should know that, as the years went by, he mellowed. He came to accept that the younger generation was not responsible for, to put it neutrally, the conduct of their forebears during the war and acted accordingly.'

Stephen did concede though, that Ray would have been 'horrified' that a car was parked across that splendid front lawn.

Chapter 2

CHARLIE MARSHALL SAYS G'DAY

IT HAS TAKEN ME 50 years but I'm finally here, standing alone in a graveyard above Anzac Cove. I look down at the headstone and whisper the words. In the silence of the morning they sound loud and rough.

'Charlie Marshall says g'day.'

AT 11 PM ON THE eve of Anzac Day, the ferry is chugging across the harbour from the Turkish port of Çanakkale to the Gallipoli Peninsula. It's bitterly cold and even though I'm rugged up with my gloves and parka, the wind cuts like a knife as I stand on the deck, staring out into the darkness. My fellow travellers are all inside, sipping coffee or haggling with peddlers selling souvenir T-shirts, but I feel obligated to experience the discomfort and try to replicate, even in the faintest way, what the original Anzacs might have felt.

That's why I'm here, to follow the footsteps of my ancestors: my grandfather Charlie and great-uncle Jock. They were two young men who heard their country's call and marched off to war — one a teenager looking for adventure and excitement, the other older and wiser. Jock's memoirs, written in 1975 when he was 85 years old, are tucked away in my rucksack. They are my spiritual road map on this short trip into the past. I have read them so many times over the past few weeks I can almost recite them by rote. They tell a tale of youthful innocence lost, of hope and horror, life and death. Of a hardworking farmer whose sense of duty is tempered by foreboding and divided loyalties.

> I was deeply disturbed because I had so much work on hand on the farm and my parents had relied on my help. It was difficult to know my duty. I had helped at home for about 9 years but I had not been asked to help my Country. After much thought and after obtaining the promise of my friend Billie Gilchrist to cultivate the remaining ploughed land and sow in turnips, I volunteered for overseas service.

By 3am we are at Anzac Cove and it is so cold that even though the bloke in the seat next to me is almost totally hidden under a pile of blankets, I can still hear his teeth chattering. He tells his mate he is trying to keep warm by imagining himself back home in Queensland in front of a fire with his dogs and a bottle of rum. I don't think it's working.

Between interviews with Gallipoli historians, Andrew Denton, host of the ABC coverage being beamed back live to Australia, keeps the crowd regularly updated with what the original Anzacs would have been doing exactly 91 years ago. How right now they would be checking and re-checking their gear or writing their last letters home; now patting each other on the back and saying a silent prayer, and now climbing into the boats and being pulled towards the headland where we sit waiting for the sun to rise.

My grandfather and great-uncle weren't in that first assault.

They were replacements, sent in later in the day.

> Sunday 25th April about 6am we left the harbour and sailed northwest towards the Dardanelles. In the distance we could hear the booming of ships and other guns. One wondered how men could live under the terrific blast of some of the big shells. At 9.30 that morning, as we sailed up the coast towards our target, later called Anzac Cove, we had Church service on the open deck. With the roar of the guns, the chatter of machine guns reaching us from shore, our thoughts on our own ordeal awaiting us, there could surely never be a more sobering or moving Church service. Our old Padre when reading the lesson 'In the midst of life we are in death' could only be heard during

> intervals between gun-blasts and from memory the two hymns sung were 'Fight the Good Fight' and 'Abide with Me'. It made a deep impression on me as it must have on all those present.

Our eyes are drawn out into the darkness, past the floodlit stage and across the beach. Out to sea where we know the ships would have been. It's almost as though we'll be able to see them if we look hard enough, those brave, naive, hopeful boys, as they lurch towards us in their little boats. In a way, we can. Finally, as the blackness turns to light and the cold night mercifully morphs into a glorious warm, clear morning, Denton tells us that right now they would have been clambering ashore and running blindly up the beach; running through the doorway of hell.

THERE ARE A REPORTED 8000 of us here on Anzac Day 2006, all with our own reasons for making the trip. I'll speak to probably two dozen of my fellow travellers during my few days in Turkey and they'll each tell me their story: about promises they made to relatives and to themselves; about wanting to see Gallipoli just once before they die, about distant uncles and family folklore, about fitting it in between Oktoberfest and the Running of the Bulls. Most of all, they'll tell me they're Australian and that meant they needed to come. They were drawn here.

Me? I'm here because of an 18-year-old boy who clambered ashore at Anzac Beach in April 1915 and somehow got off in one piece. My grandfather, Cheviot

Dundee Marshall. I was 23 years old when he died but, even so, I'd never really got to know him. I only met him a handful of times, but I knew a little about him. Like how he was named after the steamer that brought his forebears from Glasgow, Scotland, to Dunedin, New Zealand, in 1862 — the ship his father was born on. About how no one ever called him Cheviot. He was Chief to his family and Charlie to his friends. How he ran the grocery shop and raised his family in a little North Island town called Te Kauwhata. Most of all, I knew how he had falsified his age by two years and tricked his father into signing a piece of paper giving him permission to serve overseas in World War I.

I can't remember when my mother told me that story, but I do remember where. We were all in Dad's Holden, returning from a family outing. I was probably 14 or 15 and somehow the subject of Anzac Day came up. It wasn't as big as it is now. Back then, maybe a couple of hundred Australians would make the trip to Gallipoli. It was the time of the Vietnam War, of protests and peace marches, and glorification of war was decidedly uncool. At school we studied Alan Seymour's play *The One Day of the Year* and related far more closely to the rebellious Hughie Castle who scoffed at Anzac Day than to his ex-serviceman father Alf who saw it as a symbol of what it meant to be Australian. That's just the way it was in 1970s Australia, but there was still something special tucked away in our memories. In primary school we'd shaded pictures of Simpson and his donkey with our

Lakeland and Derwent coloured pencils; we'd been told that this strange battle so far away and long ago had been the real birth of our nation, and we all knew what the letters ANZAC stood for.

'My dad was there,' Mum said in the Holden, almost offhandedly. 'At Gallipoli.'

It was like a flashbulb had gone off in my face. My grandpa, a genuine war hero. Well, okay, maybe not a hero, but he had been at Gallipoli and that was something, wasn't it? Something I could grab on to and boast about on Anzac Day. Something I could put together with the vague, grey memories I had of this man I never got to know. Something with which to define him, maybe even to define me.

I pressed my mother for details, but she didn't know much at all, just that he'd been there and in France, and that his father was so angry over the forged letter that he refused to speak to him again.

I had my own memories, mostly from a family trip to New Zealand, aged six. I remembered a group of old people, gathered on my grandparents' front lawn. One man had only one arm. I was fascinated. 'He lost it in the Great War,' Chief told me when I asked later. There was the town mayor, the man they called 'The Colonel', who had a swimming pool and a display of bayonets on his lounge-room wall. And there were my grandfather's marching songs, which he would sing absentmindedly when making the early morning tea or working in the garden: *Mademoiselle from Armentieres* and the tune he taught my sister and me during those few weeks we had together, *Paddy McGinty's Goat.*

We named our family dog Paddy in his honour.

So this, then, was the jigsaw puzzle that started to come together when I was told about Chief and Gallipoli. Memories of old men, old bayonets and old songs. It would be a long time before I knew more.

When Chief died in 1978, aged 82, my parents returned from New Zealand bearing more pieces of the puzzle: a camera he had carried through the Great War; photographs of himself and fellow soldiers in Egypt prior to leaving for Gallipoli; a victory postcard he'd bought for his sister in Paris but never sent; and his certificate of service, signed by Lord Liverpool. A few years later, at the request of my mother, a plain brown envelope arrived from the Personal Archives section of the New Zealand Defence Force, Upper Hutt, New Zealand. Inside was Chief's military history.

It was all there: a full record of the four years, 231 days Lance Corporal CD Marshall, registration number 8/994 of the Otago Infantry, spent in military service, from a copy of his enlistment form with its false birthdate to his discharge papers and a record of his medals — the 1914–15 Star, British War Medal and Victory Medal — sent by mail in 1938, and the Gallipoli lapel badge and medallion, issued in 1967.

The small print, handwritten by a succession of army clerks, fills in some blanks. How he'd been part of the First New Zealand Expeditionary Force that had sailed from Wellington Harbour on October 16, 1914, and landed in

Alexandria, Egypt, six and a half weeks later. How he'd undergone training in Alexandria before embarking for the Dardanelles on April 12, 1915. Four days later he was sent to the sick bay suffering measles, but on April 24, the day before the Anzac landing, he rejoined his unit. As the entry says simply: 'Recovered and sent to front.'

And that's pretty well it. There are a couple of mentions of visits to the sick bay and on April 6, 1916, the entry 'Embarked (Egypt) for France'. Not much more information was ever forthcoming from Chief. A Kiwi cousin told me, 'He said the only thing he remembered was spending the whole time running to the latrines because he had dysentery.'

It was left to his brother-in-law, my great-uncle John McIntyre White, to paint a fuller, more chilling picture of life and death at Gallipoli. I only met Jock once, on that visit to New Zealand back in 1961, but he would make a lasting impression on a six-year-old kid. He was the one-armed man.

Unlike Chief, Jock was no adventurous kid when he set sail for war. He was 24 years old, a farmer who had worked the family property and saved enough to buy four horses and a mower, reaper, binder and double furrow plough, which he used to cultivate neighbours' land. He was smart and hardworking and he stood out among the boys he'd joined up with.

> Dick Hanley, who was inclined to drink more than he should in company, asked me to look after his

> purse for him in case he got into trouble, which I did, keeping it under my pillow on board the Steamer. After handing it over to him the next day he said it contained 75 sovereigns, all his money which he was handing over to a married sister. Regularly when we went to Wellington on leave on a Saturday night Dick would ask me to keep an eye on him as he was inclined to get into trouble.

My mother never made it to Gallipoli. She died on Australia Day, 2002. Three years later, on my 50th birthday, I make a resolution. I'm going for her. And for Chief.

ON APRIL 21, 2006, I board Qantas Flight 51 bound for Istanbul via Singapore and London. In my carry-on bag is a copy of Jock's memoirs, sent to me by his daughter Frances, and Chief's tattered, oft-stitched campaign ribbons.

At 9am on April 23, I meet up with my tour group in the car park outside Istanbul's Hagia Sophia mosque for the bus trip to Gallipoli. We'll visit the battlefields that day before heading to Çanakkale and returning the next night for the Dawn Service.

Optimistically, I've signed up with a group called Hassle Free Tours. Our guide is a 60-something Aussie named Peter Robb, who for 11 months of the year is a diesel mechanic at Port Kembla. He first came to Gallipoli six years ago and returned so many times the local tour company gave him a job. Ruddy faced, missing a front tooth and with a smoke dangling from the side of his mouth, he stands in the shadow of one of Turkey's oldest historical sites and puts it into perspective. 'I reckon it's like the Muslims going to Mecca,' Robb says. 'Every Australian and Kiwi should get here at least once in their lives.'

The first thing I notice about Gallipoli is how beautiful it is. I've heard so much about these killing fields that I'm expecting some scene of horror from an old black-and-white movie. Instead, it's a colour postcard. Noosa without the surf. I almost can't believe it all happened here. It just seems so peaceful.

> It was in this trench just beside me that one of our men was shot in the head and as he was falling our Platoon Officer caught him with his hand on the man's neck. When he removed his hand it was covered with the poor chap's brain. He was violently sick and so our experience of War evolved. In this trench beside me at the last was Dick Hanley, he of the 75 sovereigns. I gave him a leg over the back of the trench and then hopped over the crest myself. I didn't see Dick again, but was told he was shot through both legs in retreating and died later in Hospital.

Robb shows us the sights. The beach where the Anzacs were supposed to have landed, the hill where the Turkish general Mustafa Kemal Atatürk stood when an Aussie sniper took a shot at him. The mountainous slope up which the Kiwis dragged a giant water tank under enemy fire. He takes us to The Nek, where wave after wave of young Aussies were slaughtered, running into a wall of machine-gun fire to create an unnecessary diversion for a British landing on a deserted beach. Two 22-year-old girls in our group climb into a covered trench to have their photo taken. As they clamber out, Robb tells them: 'Twenty men drowned in that trench.'

And everywhere there are the gravestones, maintained by the Commonwealth War Graves Commission and lovingly tended by the Turks, who revere this land every bit as much as any Australian. Ask them and they'll tell you three nations were born here that terrible winter. It is only the British who see what they called the Dardanelles Campaign as a disaster. The Anzacs call it Gallipoli and take a day off work to celebrate. To the Turks it is the Battle of Çanakkale and theirs is the biggest service of the day, honouring the bravery and sacrifice of all who took part. The words that General Atatürk, hero of Çanakkale and the father of modern Turkey, reputedly uttered at Gallipoli on Anzac Day 1934 and repeated at every Dawn Service here since are carved in stone for all to read:

> Those heroes that shed their blood and lost their lives. You are now lying in the soil of a friendly country. Therefore rest in peace. There is no difference between the Johnnies and the Mehmets to us where

> they lie side by side now here in this country of ours. You the mothers, who sent their sons from far away countries, wipe away your tears. Your sons are now lying in our bosom and are in peace. After having lost their lives on this land, they have become our sons as well.

We see the grave of Simpson, the man with the donkey, the most famous soldier to die at Gallipoli, and that of Private Jim Martin of Hawthorn, Victoria, at 14 years of age the youngest. Walking from grave to grave, reading the heartbreaking inscriptions from parents and wives, sisters, brothers and friends, you can't help but shake your head. Just over 44,000 Allied soldiers were killed at Gallipoli and the Turks lost almost double that number. After a while, you think of them not as graves but as boys who lie on this headland waiting for visitors. Forever young.

We walk from one historic site to another. We visit the small museum and gaze up at the statue of the Turkish soldier who walked into no-man's land to carry a wounded Englishman back to his own lines. We climb to Chunuk Bair, where the entire campaign could have been won but was lost for lack of reinforcements. We see it all … all except for the little cemetery I have travelled 15,000km to visit. Quinn's Post.

'You might get a chance to get there tomorrow,' Robb says and ushers me to the bus.

He is right. The opportunity comes between the end of the Dawn Service at Anzac Cove and the start of the

Australian commemoration at 10.30am. I walk up from the beach and when the massing crowd files into Lone Pine, I keep walking up the single-lane arterial road. On the left are the trenches of the Anzacs, on the right those of the Turks, so close they could almost reach out and touch.

The sun is shining now, the sky a glorious blue.

The peaceful scene that greets me as I reach Quinn's Post could not have been more different to the mayhem of May 2, 1915, when the Otago Infantry was called up to attack. The idea was that the Aussies would go forward first, then the boys from Otago would support on their left. There is a breakdown in communication. The Kiwis are late. By the time they arrive, the Turks are waiting.

> At dawn the Turkish trenches were found to be only about 50 yards ahead of us on higher ground and they opened up intense rifle and machine-gun fire. Our people called for reinforcements on the left. I foolishly responded to the call and found myself on a rather exposed position behind our trench.

I walk from one headstone to another, reading each and moving on.

> I received a nasty bullet or bullets which practically shot away my right hand, left a ghastly wound on my left jaw and another through the left shoulder.

I pass the headstones of dozens of Australians who will always stay here at Quinn's Post, reading each in turn and moving on, still searching.

> In my youth, shooting rabbits at home, I had watched the effect of bullets or shot on a rabbit running along the hillside. They spun over several times down the slope, turning somersaults etc. This was my lot when hit on Gallipoli. I spun over several times down the slope. It was a most painful experience and soon my clothes were blood covered and I was very groggy. After crawling down the gully a bit, two men from the Naval Division helped me down the steep gulch until they found stretcher bearers to take me back to the beach. I recollect little of the journey back, but I remember trying to tell the doctor to, if possible, save my hand. My next recollection is being hoisted onto a ship out at sea by the ship's own gear. So I was finished at Gallipoli.

And then I see it. The grave of Private JC Smylie. It's not a name I've ever heard before, but the words underneath make it special. Otago Infantry.

I put my hand in my pocket and clutch the tattered tunic ribbons.

'Charlie Marshall says g'day.'

***

Postscript

Lance Corporal Cheviot Dundee Marshall was discharged from active service on March 22, 1919, and returned to Dunedin, where he bumped into his father on the street. The two men made their peace. A few years later, he moved to Te Kauwhata and in 1926 married the local schoolteacher, Vera Paterson. They had three children and ten grandchildren. Chief ran the village grocery store until his retirement in 1966.

Private John McIntyre White's right arm was amputated on the beach at Anzac Cove. He then lay on a table on a troopship for four days before his wounds were seen to. He had his 25th birthday on a hospital ship bound for England. In April 1916, he was discharged from the New Zealand Army and joined the British Army, being commissioned as a lieutenant and returning to Egypt as a transport officer. In 1926, he married Joyce Marshall, Chief's sister. They had two children and six grandchildren. Unable to continue farming, Jock worked as a salesman and was a leading advocate for the rights of returned servicemen and amputees until his death in 1978, aged 88.

Lance Corporal John Charles Smylie was the son of Mr and Mrs A Smylie of Herbert, Otago. The date of his death is listed only as sometime between May 1 and 23, 1915. He was 19 years old.

Chapter 3

THE TREE OF LIFE

IT IS 1997 WHEN I first notice the tree. With two restless youngsters in the back seat of the car and some time to kill, I turn off Waterworks Road at Ashgrove in Brisbane's inner northwest and cross the bridge that leads into St Johns Wood, an exclusive enclave of renovated pre-World War II cottages and large contemporary houses with million-dollar views of the city skyline, which is just four kilometres away. A left turn into Royal Parade and I find what I'm looking for: a playground.

As my daughters entertain themselves on the equipment, I step away and spot a large rock in front of a giant kauri pine. On the rock is a small silver plaque, held in place by four chrome-plated plugs. It reads:

TO THE MEMORY OF
FLT./SGT. CLIFFORD BERGER HOPGOOD

KILLED IN AIR OPERATIONS OVER
OCCUPIED EUROPE
24TH FEB, 1944.
THIS TREE WAS PLANTED BY HIS FRIENDS
OF ST. JOHNS WOOD.

In my imagination I see the friends and neighbours of Flight Sergeant Hopgood, some still in uniform in the days after the war, planting a sapling, stepping back and bowing their heads in silence for a few moments before walking back to their homes. I imagine Clifford as a small boy growing up nearby and playing in the very park my children are now enjoying. I see him in short pants and with Brylcreemed hair, running around with maybe a wooden toy plane. And I imagine him in his blue RAAF uniform leaving St Johns Wood and heading off to war, never to return.

There's a story here, I say to myself as I leave, vowing to one day tell it. It taunts me for over a decade, calling to me every time I drive past the turnoff to St Johns Wood. Finally, last year I go back to The Woods Memorial Playground, write down the details on the plaque and begin my search for Clifford Berger Hopgood.

In time I will find that I was at least partly correct in my musings. There was a group of Hopgood's neighbours standing around the sapling; Hopgood did leave St Johns Wood in uniform, never to return; and there was a small boy playing in the park. Just a different boy.

TODAY, IT'S A LOT easier to trace the story of a fallen airman than it would have been when I first saw the tree. The internet has seen to that. The Australian War Memorial website informs me that Flight Sergeant Clifford Berger Hopgood, official number 414565 from 460 Squadron RAAF attached to the RAF, was killed on February 24, 1944, aged 28, and buried in the village cemetery at Villers-sous-Prény, Lorraine, France. A departmental manager, he was born in Toowoomba on August 3, 1915, and lived at 89 Royal Parade, St Johns Wood. Next of kin was his wife, Margaret Warren Hopgood.

Typing Hopgood's name into Google brings up a link to the Churchie Old Boys' Association's 'List of the Fallen', honouring old boys of Brisbane Church of England Grammar School who were killed in action in all conflicts from World War I.

Churchie archivist James Mason provides me with the details of Hopgood's time at the school, where he was student number 683. He enrolled in 1928 as a 'choir scholar', meaning that on Sundays he sang in the St John's Cathedral choir, earning a reduction in fees. A talented athlete — a report in the *Brisbane Courier* of Saturday, September 8, 1928, on Churchie's 14th annual sports day notes that J. Javes won the junior cross-country 'after a hard tussle from Hopgood and Nixon-Smith' — he competed in the school's annual Pocket Swim around Norman Creek and scored tries in rugby matches against Nudgee and Brisbane Grammar in the 1930 season.

Hopgood left Churchie at the end of 1931 after sitting his Junior Examination and achieving good B Level results in English, Latin, Arithmetic, Algebra and Geometry.

As we speak, Mason's fingers tap on his computer keyboard. 'Have you tried the National Archives?' he asks. 'Here it is … 79 pages.'

Those pages provide a poignant and heart-wrenching picture of the days and nights that followed the take-off of Hopgood's plane, Lancaster bomber ND394, from RAF Binbrook airfield, near Bookenby, Lancashire, at 6:02 on that night in 1944. On board was a crew of seven that included Australians Warrant Officer DW Baxter (pilot), Flight Officer A D'Arcey (bomb aimer), Flight Sergeant RC Ferguson (wireless operator) and Warrant Officer WJ Martin (upper gunner). Also on board were the RAF's Sergeant P Mallon (flight engineer) and Sergeant JD Dunlop (rear gunner). Clifford Hopgood was navigator. The mission was to bomb factories in Schweinfurt, Germany.

According to the Australian War Memorial's official history, 'Nothing was heard from the aircraft after take-off and it did not return to base; 17 aircraft from the squadron took part in the raid and two, including ND 394, failed to return.'

It was four days before Margaret Hopgood received notification of her husband's presumed fate in the form of a telegram from RAAF headquarters in Melbourne. A copy, its edges torn and frayed, is held in the National Archives. It reads:

> PERSONAL Mrs CB Hopgood. Regret to inform you that your husband 414565 Flight Sgt Clifford Berger Hopgood is missing as result of air operations STOP Details are he was member of crew Lancaster aircraft detailed to attack Schweinfurt Germany which failed to return to base presumably due to enemy action STOP The Minister for Air joins with Air Board in expressing sincere sympathy in your anxiety STOP When any further information is received it will be conveyed to you immediately.

The telegram was followed on April 6 by a letter from Wing Commander HD Marsh, DFC, Officer Commanding 460 Squadron.

> Dear Mrs Hopgood, I write to offer you the sincere sympathy of myself and the Squadron, during the anxious time you will be passing through. Your husband was very popular in the Squadron and had carried out his duties in an extremely conscientious manner. It may be of some consolation to know that a number of missing airmen are eventually found to be Prisoners of War and I hope I will have the pleasure of passing such good news on to you before long.

There would be no such good news. Not for Margaret Hopgood, anyway. On June 17, 1944, she received a second telegram from Australian officials:

> Information received from International Red Cross that one Australian member of your husband's crew Warrant Officer WJ Martin is a Prisoner of War STOP Regret no news of your husband or remainder of crew.

Six months later, Margaret wrote to MC Langslow, secretary of the Air Department, care of the RAAF's casualty section in Little Collins Street, Melbourne — an office that over years of correspondence would be Margaret's source of early hope and ultimate sadness. Margaret's neat, handwritten letter states: 'It is with regret I inform you that I have had no news from him since he was reported missing.' Then, on February 15, 1945, almost a year after her husband's fatal flight, Margaret Hopgood received in the post a letter signed on behalf of Langslow.

> It is with deep regret that I have to inform you that the death of your husband has now been presumed for official purposes to have occurred on the 24th February, 1944. The Minister for Air and members of the Air Board desire me to extend to you their profound sympathy in your great loss.

Margaret was sent a list of the personal effects of 'the late 414565 F/Sgt Hopgood C. B.' Among such items as 18 handkerchiefs, 24 socks and a leather tobacco pouch, was 'one wrist watch on leather strap, engraved "To Cliff from Margaret"'.

In November 1944, Margaret had been informed by Langslow's office that two of her husband's crew, pilot Baxter and bomb aimer D'Arcey, had escaped through occupied France and made their way back to England. The crewmen reported that the aircraft had been attacked by a night-fighter at 22,000 feet. The starboard engine caught fire and pilot Baxter instructed the crew to bail out. D'Arcey reported getting out with Hopgood at 15,000 feet but had 'no further knowledge' of him or the rest of the crew.

It would not be until the months after the war that Margaret would learn about her husband's final moments. In September 1945, Langslow's office wrote that Flight Sergeant Martin, recently repatriated from a prisoner-of-war camp, 'was told by a German interrogation officer that your husband was found lying in a recumbent position in a field without any sign of external injuries'. The report continued: 'Further examination showed that his neck had been broken, apparently in landing as his parachute was open.'

Eight months later came another letter signed on behalf of Langslow. It was the last that Margaret would receive from his office. It read: 'A copy of a letter from Madame Schuller of Villers-Sous-Prény, addressed to the British Ambassador in Paris, has been received. As it is thought that the information may be of some comfort to you in your sorrow, a copy of the letter is enclosed.'

Written on October 21, 1945, and translated by the office of the British ambassador, Madame Schuller's letter tells of the night the Lancasters passed over Villers-sous-Prény. It was 10.25pm and 'unfortunately the Luftwaffe were very active and hit your bomber in full flight'. Villagers hurried to the crash scene and found a badly wounded Flight Sergeant Martin. Madame Schuller's husband, the local baker and grocer and for four years a member of the French Resistance, told the villagers to bring Martin to their house, where the town doctor treated his wounds. Baxter and D'Arcey were brought to the house and hidden. With Martin needing urgent surgery, the Schullers reluctantly notified the Germans, who interrogated him and took him away. Baxter and D'Arcey, who remained undetected despite a search of the house, were given false papers and assisted passage to England.

But it was the second part of the letter that resonated most with Margaret Hopgood:

> The following four airmen are buried in our little cemetery: Hopgood, Dunlop, Mallon and Ferguson. My husband identified them by the names on their parachutes. I tell you straight away that they were killed instantaneously. The Mayor told us to pick up the dead — four young men cut off in the prime of life by the bloody war. But they had a funeral worthy of your glorious soldiers, as was proved by the long procession of mourners who attended the ceremony — there were more than 6000 people taking part in the funeral. You can imagine how furious the Germans were when they realised what

> had happened. There were Gestapo reports and interviews — these included my husband who brought the flowers and wreaths — and ramifications followed. The Mayor was fined 3000 francs for having allowed the ceremony to take place in his district and they threatened to take hostages.
>
> Tell their mothers that although there is only one grave, each man had his own coffin. They lie in our little village and their grave is decorated with flowers and wreaths and always has been, in spite of the Germans. Tell their mothers also that the French fully realise what this loss must mean to them … I also had a son, who now lies side by side with his comrades.

And that is where my story of Clifford Berger Hopgood would have ended, if not for another instance of his name in Google results. As part of the national Avenues of Honour Project, which is researching and documenting all war memorials as part of the centenary of Anzac commemoration in 2015, Brisbane City Council has produced a report listing and describing all memorials within its boundaries. The report, which appears online, includes a photograph of the tree, the rock and the plaque. Underneath are the words, 'A local resident, Marjorie Summerville, recalls being present at the planting of the memorial tree …'

'YES, I REMEMBER BEING THERE,' 89-year-old Mrs Summerville tells me over the phone. 'I'm not sure how many were there. Everyone who lived near the park. George Colville organised it. He was a soldier too. I never knew Cliff Hopgood. We moved there when my husband got back from the war in 1946 and the tree was planted not long after that. It was directly opposite their house. We were at 101 Royal Parade and when we came, there weren't any houses in between us and them. I used to go down and use their phone; we didn't have one then.'

We speak for a while as Mrs Summerville recalls the day they planted the tree. I ask if she knows what happened to Margaret. 'Oh, she died,' she says. 'The house is gone too. They cut it in half and took it away on a truck.'

She then adds, 'You know they had a son, don't you? Robert. He lives on one of the islands.'

ROBERT HOPGOOD IS WAITING for me at the jetty as the ferry pulls into Macleay Island. A fit looking, energetic 70-year-old, he takes me for a brief tour of the island in his four-wheel-drive.

'Would you like to have a look at this?' he asks as we pull up to a small but striking building. It's the 100-seat non-denominational Macleay Island Church of the Holy Spirit. Robert leads the way inside.

'When we came here 14 years ago there was no church on the island, so I designed this and we built it with volunteer labour,' he says. 'The thing is, I never knew my father, so I can't really tell you what he was like, but I thought maybe if you saw this place it might say something about him, and about my mother too. The person you become … it has to come from somewhere, doesn't it?'

A few minutes later, I'm sharing a coffee with Robert and Alison, 68, his wife of 48 years, in the large eat-in kitchen of their waterfront home. We are seated on chairs manufactured by Harley Furniture, a division of the successful home-building, earth-moving and construction company Robert founded after graduating from the University of Queensland with a degree in architecture in 1963. The 180-degree view north across Moreton Bay is spectacular. The Hopgoods' 13-metre catamaran *Eaglehawk* is moored nearby.

On the table before us are the few mementoes Robert has of his father: his parents' wedding photo; the framed picture of Clifford in his airforce uniform that stood on Margaret's bedside table until the day she died; and 'one wrist watch on leather strap, engraved "To Cliff from Margaret"'.

One of Robert's earliest memories is the day his mother took him — aged five or six — on the tram to South Brisbane railway station. They sat silently on the platform waiting as a train pulled in. 'Two servicemen in uniform got off and came up to us,' he says. 'One of them gave me a little wooden toy. It was a penguin on wheels, with a couple of bits of leather that made a noise when you pushed it along. Then they gave something to Mum. I remember pushing this little toy and wanting

Mum to look, and then I saw she was crying. It was the watch. They'd given her Dad's watch.'

When Robert turned 13, Margaret gave him the watch. He wore it until it stopped working a decade later. 'I've never worn a watch since,' he says.

The watch had been Margaret's engagement present to Clifford. They had met in the mid-1930s. She was working at O'Brien's Paper Bag factory at inner-city Spring Hill; he was sales manager at Holeproof, the hosiery company. Born in Charters Towers in 1911, Margaret moved to Spring Hill in Brisbane with her father, two sisters and baby brother after her mother died in childbirth. Her oldest sister Sarah, 11 at the time, left school to help raise the younger children. The family struggled during the Depression, to the point where Margaret refused to marry Clifford until they had built their own home.

'As a young girl, she'd been the one sent to tell the landlord that they couldn't pay the rent,' says Robert. 'She never wanted to go through that again.'

The couple bought land in Royal Parade, St Johns Wood, and for several years their outings involved picnicking on the site as they saved to build the home, which they named *Glen Arran*. They married on May 27, 1939, just over three months before Australia entered the war against Germany.

Robert was born on September 27, 1940, with Clifford promising Margaret he would not leave for the war until they had shared their son's first birthday. On October

12, 1941, Clifford left his office in the city and went to the Number 3 RAAF Recruiting Centre nearby to enlist, taking an oath to 'resist His Majesty's enemies and cause His Majesty's Peace to be kept and maintained, and that I will in all matters appertaining to my service faithfully discharge my duty according to law. So help me God.'

Following aircrew training in NSW, Clifford left Melbourne by troopship for Canada on November 2, 1942. Margaret rented out *Glen Arran* and took Robert back to live with her family in Spring Hill. Clifford arrived in England in early 1943. After completing a navigators' course and serving with RAF 625 Squadron, he joined RAAF 460 Squadron on November 28, 1943. Three months later he flew his final mission. There were eight men who left St Johns Wood to serve in World War II. Clifford Hopgood was the only one who didn't come home.

By the time Margaret and young Robert returned to the family house in St Johns Wood in 1948, the tree was already there. George Colville had asked council for permission to plant it and they had supplied the sapling. A white timber and chain fence was constructed around it and the plaque affixed to a post. As soon as he was old enough, Robert would clear the long grass around the memorial, first with a reaping hook and then a Qualcast mower that Margaret had purchased from a neighbour. Years later, when the fence was damaged by a council tractor, a giant granite boulder was brought down from Mt Coot-tha and the plaque attached. For Robert, who has always referred to it as 'Dad's tree', the healthy kauri was a constant reminder of what he was missing. In many ways, the only reminder.

> The way that Mum and her generation dealt with grief was that there was no counselling. You just got on with your life and tried to lock out anything that hurt you. I learnt that at a very early age. If I'd said, 'Why don't I have a dad?' — or something like that — I'd have been told to eat my dinner. It was, 'We don't talk about that.' The pain was completely washed out of me. I was like, 'Yeah, I had a father, he was killed in the war, so what?' It didn't mean a thing.
>
> At the same time, the influence of Mum on my life, things like loyalty and work ethic, was quite unreal. It was hard for Mum. She became very independent, she needed to be, but at a very hard time in terms of social convention. She liked to take me away on holiday, but a woman sitting alone on a beach in those days was just not done.

Margaret returned to work at the paper bag factory and remained there until her retirement in 1971. She also became a stalwart of the War Widows Guild, representing the organisation at Anzac Day ceremonies. For young Robert, they weren't enjoyable experiences.

> I went to as many Anzac Day services as anyone. I approached them with trepidation. As a war widow,

> Mum was expected to go back to the RSL club after the service. There would be all these drunk blokes and this young woman and it was very uncomfortable. I remember being upset and Mum saying to me, 'Robert, Anzac Day is that day of the year when those who've returned from the war remember those who didn't … but we remember them *every* day.

In 1975, Margaret accompanied a group of 27 war widows on a tour of France. She had made arrangements to slip away on her own to visit Clifford's grave in Villers-sous-Prény, but when her companions heard of her plans, many of them went with her.

When their bus pulled into the village, they found the footpaths lined with people. A banner strung across the road proclaimed: 'In Memory of Your Dear Deceased. Australian Friends, You Are Welcome in Our Village.'

A public holiday had been proclaimed in their honour and a civic reception held in the town hall following a march from the site of the plane crash. As Margaret stepped from the bus, there to meet her were the mayor who had been fined 3000 francs for authorising Clifford's funeral, the doctor who had attended Flight Sergeant Martin, and Madame Schuller.

A year after Margaret's death in 2008, her daughter-in-law Alison persuaded Robert to also make the pilgrimage to Villers-sous-Prény. From the day Robert first brought Alison home to meet his mother in 1958, she and Margaret had formed the closest of bonds. An accomplished choral singer, Alison performed with the Brisbane Eisteddfod Junior Choir and Margaret was her No. 1 supporter, attending all concerts.

'I wasn't all that keen to go to France but Alison talked me into it,' Robert says. She adds, 'I told him if you don't come, I'll go by myself.'

The Villers-sous-Prény that Robert and Alison walked into in 2009 was a different village to the one that had welcomed Margaret 32 years earlier. The streets were deserted, the shops closed. 'There's no work around,' Robert says. 'The whole place was empty.'

They found the church and Robert stood before his father's grave. 'Basically I was just standing there and thinking, *This is him. I really do have a dad.* For the first time, I felt like I had a father.'

He placed some of his mother's ashes on the grave, then he and Alison tried without success to gain entry to the church. 'It was locked up, but just by chance I saw an old woman in a garden nearby,' Robert remembers. 'I somehow managed to explain what we wanted and she went to another house and a woman came out with a big old key and unlocked the front door for us. It was a lovely little church, beautifully looked after.'

As Robert sat in a back pew, Alison stood at the front of the church and sang Margaret's favourite song, *These Are The Lovely Things*.

In her latter years, Margaret Hopgood would sit quietly for hours on the veranda of the home she and her husband had built, looking across to the park at Clifford's tree, whose roots spread all the way to France. 'I'm not going to die,' she'd say, 'until it's the tallest tree in the park.'

Margaret passed away, aged 96, on July 25, 2008, with Robert and Alison at her bedside and their children Catherine, Clifford and David close by. A modern bungalow now stands at 89 Royal Parade. Two ageing corner fence posts are the only reminder of *Glen Arran*, the house Clifford Berger Hopgood left to go to war.

But the kauri pine grows strong at St Johns Wood. It is the tallest tree in the park.

Chapter 4

CREW

'TREE OF LIFE', THE story behind the tree planted in memory of Cliff Hopgood, was published in *Qweekend* magazine on the Anzac Day weekend of 2011. Seven months later, it won the Walkley Award for best magazine feature of the year. Equally importantly, it was read by Gold Coast resident Gabrielle D'Arcey, whose father Tony had also been part of the crew of the Lancaster bomber shot down over France in World War II. Tony D'Arcey had left behind a wartime diary. Gabby asked if I would like to read what her father had written. When I replied very much in the affirmative, she sent me a copy.

The article I had written about Cliff Hopgood and the people of the small French village of Villers-sous-Prény, in northeastern France, who risked their lives to give him and three of his crewmates a funeral attended by thousands in defiance of the Germans had, until then, been the story of one man and one family.

I had spoken to Cliff's son Robert, who had no memory of his father, but told me of the enormous influence his mother Margaret had had on his life as she raised him alone. He had shown me the picture of Cliff in uniform that had sat beside Margaret's bed until the day she died aged 96 in 2008, and the wristwatch, inscribed 'To Cliff from Margaret', that she had given him on their engagement.

I spoke to one neighbour who was there when the memorial tree was planted across the road from the Hopgoods' home in 1947, and another who had arranged for the giant boulder to be brought down from Mt Coot-tha and placed in front of it.

Robert told me about Margaret's visit to Villers-sous-Prény in 1975, when the whole town was waiting to greet her as she stepped off the bus. And about his own visit along with his wife Alison in 2009, when he stood above Cliff's grave and thought, *This is him; I really do have a dad.* He had told me how he'd scattered some of Margaret's ashes on the grave and how they had gone into the tiny, empty church and Alison, a chorister, had sung Margaret's favourite song, *These Are The Lovely Things.*

That, I thought, was the story of the tree. The story of the Hopgoods. And then I read Tony D'Arcey's diary.

THERE WERE SIX OTHER young men aboard the RAAF Lancaster bomber J For Jig when it took off from Binbrook air base in Lincolnshire, England, on February 24, 1944. They were on an operation to bomb factories manufacturing armament components in Schweinfurt, 160km east of Frankfurt in Germany. I had seen their names on official documents when researching my original story: Australians David Baxter, the pilot, 21, from Wonthaggi in country Victoria; Anthony D'Arcey, 20, the bomb aimer from Sydney; Ronald Ferguson, the 27-year-old wireless operator from a property near Bundaberg; William Martin, 32, the mid-upper gunner from Sydney; Scotsmen Peter Mallon, 21, the flight engineer, and John Dunlop, 19, the rear gunner.

As I wrote my *Qweekend* story about Cliff and Margaret and Robert and the tree, the others were just that: names; shadowy minor characters in the narrative. Tony's words changed that. He wrote about who they were, where they came from, and gave funny little descriptions of their personalities and nicknames. He brought them out of the shadows and into the light. It was almost as if he was challenging me: 'What about us? Don't we matter too?'

I decided I had to tell all their stories. As I wrote in the foreword to my book *Crew — The Story of the Men Who Flew RAAF Lancaster J For Jig* …

> There were seven men aboard that flight — seven out of a total of 125,000 who served as air crew for RAF Bomber Command in England between 1940 and 1945. Their backgrounds were not unusual. They weren't a special crew, a famous crew; they were as ordinary as can be. And that's what makes them important. Because their stories are also the stories of

> the 125,000 — who they were, what they did, whom they loved, and whom they left behind.

With publishers Allen & Unwin giving the project the green light, my next call was to Maureen D'Arcey, Tony's widow. Maureen, who lives on the NSW Central Coast, told me that pilot David Baxter's widow, Joyce, had moved from her home at Wonthaggi into a nursing home several years before. She didn't know the name of the nursing home, wasn't sure if Joyce was still alive, and the only phone number she had was the old one from Wonthaggi. With nothing else to go on, I gave it a call.

'Hello?' a frail voice answered.

'Is that Joyce?' I asked.

'Yes …' she said. Incredibly, Joyce's family had transferred her home phone number to the nursing home, and another piece in a jigsaw puzzle it would take me six years to complete fell into place. Joyce told me I should ring her son Alan in Melbourne. He'd know everything about his father, she said — and he did.

There was another lead that Maureen D'Arcey gave me during that initial conversation. She told me that gunner Bill Martin and his wife Anne were no longer alive, but they'd had two children, Richard and Christine. 'Christine trained to be a nun,' she told me, but other than that she didn't know anything about them. We spoke for a while longer and I was just about to end the conversation when she said, 'Oh, wait on, I seem to remember …'

On the top right-hand corner of a page of notes I took that day is scribbled, 'Sacred Heart Convent, Kensington?' Just on spec, I sent an email to the convent in Sydney. I was trying to track down someone who might have been a novice there, maybe 50 years before, I wrote. Her name was Christine Martin. Her father Bill had been in the RAAF …

I didn't hear back from the convent, but a month later, maybe two, I received an email from South Africa. 'Dear Mike,' it said. 'I have been tracked down. I am a missionary in Tzaneen.' It was signed 'Sister Christine Martin'.

And so it went on, like pulling on a loose piece of thread, each gentle tug unravelling the story of the four who were killed and the three survivors.

Gabrielle D'Arcey wasn't the only person with a connection to J For Jig who had read the *Qweekend* article. Armand Casalini, 77, was four years old when the burning aircraft crashed in a field near his parents' home in the centre of Villers-sous-Prény. He remembers being awoken by an explosion that shook the house and shattered the windows; how his mother Lucie had run into the room and comforted him and his seven-year-old brother Noel. He remembered hearing the stories, told over and over, of how the villagers went to the field and found four bodies and one badly injured 'aviateur'. Of how they carried him to the home of Eugene Schuller, the baker, and, most of all, how they had stood up to the Germans and had given the four comrades a heroes' funeral.

In the 70 years that followed, Armand had become the keeper of the flame that sustained the legend of the night the war came to his village. He collected parts of the plane's wreckage, preserved photographs of the crash site and funeral, and even kept the US Air Force bombardier's bag in which Cliff Hopgood carried his maps. Armand had found the *Qweekend* story online, managed to have it translated, and made contact. Another piece of the jigsaw fell into place.

I spoke to brothers, sons, daughters, nieces, nephews and friends, and with every conversation I learnt more about these seven young men who had left their peaceful lives and entered a parallel world so alien and terrifying that at times I struggled to comprehend it. Never more so than in 2012 when my wife Linda and I travelled through England and France to further research the book. Like Robert Hopgood, we stood above the communal grave with the four headstones in the churchyard at Villers-sous-Prény and felt the tears well at the senseless waste of it all, but it was at a former Bomber Command base at East Kirkby in Lincolnshire that the day-to-day reality of that alien world truly hit home.

Not far from 460 Squadron RAAF base at Binbrook where the crew of J For Jig were based, the Lincolnshire Aviation Heritage Centre, with its meticulously restored barracks, mess hall and control tower, boasts England's only Lancaster bomber still housed on an airfield. For half an hour, with curator Andrew Panton as my guide, I had the opportunity to clamber through that aircraft. I sat behind the controls, just as J For Jig's skipper had with his flight engineer by his side. I looked up into the cramped upper gun turret and eased my way past the navigator's table and radio equipment. I climbed down into the nose and looked through the perspex cover, just as the bomb-aimer would have, watching tracer bullets shooting upwards towards him from the burning target area. Most mind-blowing of all, I pushed, dragged and squeezed myself into the death-trap that was the rear-gunner's pod. It was in that claustrophobic, freezing, isolated position that the young volunteer known as 'Tail-End Charlie' would sit, hour after hour, wondering how long his luck could hold before a Luftwaffe night-fighter or German ack-ack gunner got the aircraft in their sights, knowing that, once hit, he had almost no chance of getting out alive. After extricating myself and making my way back to the centre of the aircraft I looked around at the flimsy canvas walls and awkward escape routes and asked myself, *How? How could they have done this, night after night?* And more crucially, *Could I have done it?*

On that same trip Linda and I visited the Bomber Command Memorial in Green Park, London, opened a few months earlier by Queen Elizabeth. The centrepiece is a stunning 2.75m-high bronze sculpture of a Lancaster crew, immortalised as if standing beside their aircraft having just returned from a mission. In my mind, as I looked at the superbly sculpted figures in their flight gear, they weren't just any crew; they were my crew, the crew of J For Jig.

In the years that followed, as I researched and wrote, picking up a scrap of information here, grasping gratefully at a helping hand there, I imagined the seven of them standing behind me, sometimes nodding in approval, other times urging me to keep going. At times it wasn't easy. Have you ever wondered how many Fergusons there are listed in the Brisbane online phone book? I don't know the exact number, but it's a lot, and I must have rung all of them before I finally found Ron Ferguson's niece, Diane Ellem.

The hardest crew member to trace was young Scot John Dunlop. British military archives aren't as forthcoming as their Australian counterparts. In desperation, I placed a pleading article in a Scottish historical newsletter. A gentleman named Jim Blackley took pity on me and helped me track down John's family in Canada. The last piece of the puzzle. Well, almost the last.

Dave Baxter's brother Ron, now 83, travelled to Villers-sous-Prény, where he met Armand Casalini, who gave him Cliff Hopgood's map bag. A few weeks later, I was at the tree at St Johns Wood when Ron presented the bag to Cliff's son Robert.

It was from there, beside the tree, that I had started my journey 16 years earlier, and it was with the tree that I ended it. Along the way I learnt plenty. About wartime history, of course; I now know an awful lot about various types of military aircraft, and how and where the different members of a Lancaster bomber crew were trained. I know about Pathfinders and Target Indicators and the deadly Luftwaffe tactic of *Schräge Musik*. I can tell you the difference between a Mae West and an Irvin jacket, and recite several lines of Churchill's best speeches.

But all of that you can find in a book or online. What the crew of J For Jig taught me was far more important. Far more personal. They taught me the true meaning of words that had rolled off my tongue or keyboard all my life without a second thought. Words that personify a generation that is now all but gone. Words such as duty, sacrifice, selflessness, discipline, and honour.

The last sentence I wrote before I sent *Crew* to the publisher was the ending of the acknowledgments section. After thanking the family members of the seven men who flew in J For Jig for allowing me to write about their loved ones, I wrote: 'I hope I have done them justice.'

Them, and all of the 125,000.

Chapter 5

OPERATION MONTY

IT IS MARCH 11, 2010, and Australia's most highly decorated Vietnam War veteran, Keith Payne, is pushing his way through towering bamboo and dense undergrowth in an isolated part of Vietnam's Kontum Province, 11km from the Vietnam-Cambodia border. He is wearing a long-sleeved shirt and trousers as protection from insects and sharp branches, but still his right hand is cut and bleeding. The heat is just overwhelming — 'smoking hot', as one of the US Army Intelligence officers accompanying him describes it — but 76-year-old Payne refuses to slow down or take a break. He can't. It has taken him half a lifetime to get here and time is fast running out.

He hears his voice, not certain if it is in his head or if he is speaking out loud.

'Come on, Monty,' he says. 'Talk to me, mate. Where are you?'

A few metres away, Gerard 'Jerry' Dellwo, a 63-year-old postal worker from Spokane, Washington, is pushing the thick foliage away with a stick, hoping, praying, to find something, anything, that will finally put his mind to rest. He has replayed the horror trek that he, Payne and their dying comrade took over and over in his mind a thousand times. He could have sworn he knew exactly where they had trudged that night; that he could go straight back to the place where they had left the body. But everything has changed. And it was so long ago.

The last time Payne and Dellwo were on this ridge was the night of May 24, 1969. Then, like now, Payne led the way. A 35-year-old father of five originally from Ingham, North Queensland, he was commanding the 212th company of the 1st Mobile Strike Force Battalion, a unit of 85 inexperienced indigenous Vietnamese known as Montagnards. Recruited from local mountain tribes, nearly half the company had never been under fire before and had received just 12 days of training. They were seconded to US Special Forces (the 'Green Berets') and Warrant Officer Payne's orders were to leave their defensive position about 2pm, climb the next hill where a North Vietnamese Army force was thought to be positioned, and engage the enemy. A few paces behind him as he moved out was his medic, 22-year-old Sergeant Jerry Dellwo; to their right was another unit commanded by Sergeant Anastacio 'Monty' Montez, 39, from Presidio, Texas.

About 250 Australian, US and Montagnard soldiers headed up the hill that afternoon. When Payne gave the order to pull back four hours later, there were fewer than 100 left. They had walked into a classic annihilation ambush set by as many as 5000 crack North Vietnamese Army troops. Payne got out of the ferocious encounter with wounds to his head and hand, and a commendation that would see him awarded the Victoria Cross. Dellwo was hit in the chest with shrapnel and almost drowned in his own blood. Montez didn't get out alive. For the two survivors, the physical wounds have long since mended but painful memories remain, along with four decades of 'if onlys'.

It is those doubts and regrets that the two men now have the chance to confront once and for all. For two days they will scour the one-time killing field for a sign that might lead them to Montez's body. For 41 years they have kept their memories to themselves, not even sharing them with each other. This is their chance for closure.

The two old soldiers have been brought to Kontum by the US Army's Joint POW-MIA Accounting Command (JPAC), the Hawaii-based task force whose motto 'Until They Are Home' says it all. JPAC was formed in 2003, its mission to account for — and, if possible, retrieve — the more than 40,000 US military personnel deemed recoverable in conflicts from World War II onwards. Montez is one of 1720 Americans still unaccounted for in Vietnam. JPAC has 18 teams that travel the world carrying out the meticulous detective work that can result in the recovery of remains that are then sent back to Hawaii for possible identification. The team accompanied by Payne and Dellwo is led by Warrant Officer Lawrence Renas.

A 41-year-old Intelligence officer on his eighth mission for JPAC, Renas is in awe of the way the two older men have stuck to their task despite the heat and difficult terrain. It won't be until they are packing up to head home that Dellwo reveals he tore the meniscus in his knee a few weeks earlier. Keeping the injury to himself for fear JPAC would not let him make the trip, he straps on a knee brace each morning and pushes through the pain.

'They are amazing,' Renas says. 'You can see in their faces how much they want to find this guy. I feel for them because we need it to be right. To go to the next step, we need more than just an X on a map. The conditions are tough. I know how hard it is for me, but these guys just won't stop.'

In the years since they were last here, the forest has been cleared by commercial loggers and a sea of bamboo has grown up in its place. The distance between where the battle took place and the area where they were forced to leave their comrade's body seems ridiculously short.

Dellwo looks from one point to the other. 'Keith,' he says. 'You mean this little bit of ground took us that long to travel?'

Payne nods. It did. Every tortuous centimetre.

IT WAS JUST GETTING dark when Payne decided to withdraw from the hill. Many of the Montagnards under his command who had not been killed or wounded had already dropped their weapons and run. A piece of shrapnel tore through the housing of Payne's M16 carbine. He threw it aside and picked up another that had been discarded by one of his men. A tree nearby was hit by mortar fire and a huge splinter embedded itself in Payne's scalp. To his right, Monty's company was taking an equal pounding. Payne began to pull back, throwing grenades and firing as he went.

'Get back!' he shouted. 'Get off the hill!'

Dellwo heard the call and was starting to move when he spotted two men coming towards him. It was fellow medic Paul Auriemma, half-carrying, half-dragging Montez, whose jaw had been shot off, his face horribly mutilated.

Together the two young Americans helped Montez to the bottom of the hill before Dellwo climbed back up to retrieve his gear. After grabbing his first-aid bag and two radio backpacks, he started to struggle down the hill but, weighed down with the equipment, plus his rifle and pistol belt, he tripped and fell. The strap of his first-aid bag caught on a tree branch and as he slipped and rolled down the hill, the bag was ripped from his grasp. He was climbing back up to get it when he heard gunfire and enemy voices nearby. He froze and turned back. Every day since, he has regretted that moment. If only he had kept hold of the first-aid bag, he might have been able to save Montez. If only …

Halfway down, Dellwo met Payne, who was climbing up to look for him. They caught up with Auriemma and helped carry Montez 200 metres around the base of the hill to where the others had gathered. A pathetic sight greeted them. Stunned Montagnards, most having discarded their weapons and equipment, sat in the long grass. The wounded lay moaning and crying in pain. A young, inexperienced American lieutenant who had joined the operation only that morning was on a radio calling for a napalm attack on the hill as Payne approached.

'Countermand that order,' the Australian snapped. 'We've still got men up there.'

Grabbing a radio, he said: 'I'm going back up to get them.'

Dellwo wanted to go back for his aid bag, but Payne ordered him to stay with the wounded and kept walking. Dellwo watched him disappear into the darkening night, convinced he would never see him again.

With Payne gone and Montez wounded, the lieutenant weighed up his options. After ordering the able-bodied men to move the wounded another 40 metres around the base of the hill and further into the undergrowth, he came to a decision. 'Dellwo, Auriemma,' he said. 'Come on, let's go. We're moving out.'

What Dellwo knew, but the lieutenant didn't, was that Auriemma had gone into shock. And, order or no order, Dellwo was not going to leave Montez and the other wounded behind.

'I'm staying,' he said. Furious, the lieutenant turned on his heel and started back to the previous night's defensive position, taking the able-bodied men with him. Dellwo was left with five wounded men, two of whom were in a critical condition, a delirious Auriemma, a broken radio, one rifle and next to no ammunition. He did the best he could for the wounded, organised them into a tight circle and lay down with the rifle across his chest, waiting in the darkness for the inevitable arrival of the North Vietnamese and almost certain death.

What happened next is well known to students of Australian military history. It is written in the commendations that led to Payne being awarded the Victoria Cross and the US Distinguished Service Cross, the highest military honour that can be bestowed on a non-American citizen. Three times he went back up that hill alone in the darkness, retrieving 40 of his men from under the noses of the enemy and leading them to safety. Less well known is the story of Monty.

When Payne brought his motley band of Montagnards back to the spot where he had left Dellwo, Montez and the others three hours earlier, they were gone. Tracking them using bush skills he had learnt growing up in Ingham, he came across Dellwo and the wounded hiding in the scrub. Thinking the noise he heard in the pitch black was the enemy, Dellwo clutched his rifle and thought, *This is it.* He then heard one of the wounded say, 'Officer Payne.'

'I was ecstatic,' Dellwo recalled for me in an interview nearly 40 years later. 'For the first time I thought we were going to be all right. I trusted Keith. I knew we were in good hands. All of us were so happy; we couldn't believe he had found us.'

Payne told Dellwo to get the wounded up for the trek back to the defensive position. Dellwo motioned towards Auriemma, who lay nearby, mumbling incoherently.

Payne went over and pulled him to his feet. 'Paul,' he said forcefully. 'We're in trouble here, mate, and I need you. So pull yourself together and let's get going.'

Auriemma snapped out of his semi-delirious state and helped Dellwo with Montez. They started out, Payne's Montagnards helping the other wounded.

The commendation for Payne's VC describes the next few hours thus: 'Personally assisting the seriously wounded American adviser, he led the group through the enemy to the safety of his battalion base.'

The account Dellwo provided two years ago for my book *Payne VC* comes closer to capturing the trauma and emotion of that trek.

> It was very slow and Montez was a very heavy man. Keith was on the radio most of the time, trying to get through to someone, but he'd help with Montez as well. Some of the time Paul and I were carrying Montez on our backs. I don't know how long we were going but it seemed like hours. We were going up hills and down knolls and valleys. It was dense forest, with thick underbrush and briars. The ground underfoot was slippery, damp clay and we were sliding around, but we just kept dragging ourselves through the jungle.

As they pushed their way through the undergrowth, exhausted and almost in a trance, Dellwo felt something warm and sticky on his hand.

> I was dragging Montez along with his head on my shoulder, and I felt a tapping sensation on the tips of my fingers. I realised it was blood from his jaw and that he had been bleeding a long time and was getting weaker. I called Keith over and told him we couldn't move Montez any more. Keith agreed to stop and while I tried to look after Montez, he kept trying to make contact on the radio. Finally, he made contact with a plane overhead and it relayed Keith back to headquarters at Dak To.

The commanding officer at Dak To ordered a medical team into his personal chopper with a rope and harness to pull Montez from the ground. Payne's group found a clearing and waited, listening for the sound of the helicopter or the noise of the enemy approaching.

> It was then Keith had to make one of the most difficult decisions of his life. He had over 50 men to worry about, plus the chopper could be shot down and it was unlikely Montez was going to make it anyway. He just said, 'No, this is crazy, we're not doing it. Come on, we're going up the hill', and he started leading the way. We hadn't gone 20 metres when Montez sort of gasped and collapsed and stopped breathing. I called

> to Keith and tried to give Montez mouth-to-mouth but he'd been shot up so bad, there was no mouth. I said, 'I've got to do a tracheotomy,' and asked if anyone had anything sharp. Keith had a knife and it worked and I managed to get some air into his lungs, but Paul said, 'There's no pulse,' and started external heart massage. But that was it, he was gone.
>
> Keith was right. Monty never would have survived the harness. I was crying and cradling his head in my arms. Through everything that had happened through that day and night we had all stayed so strong and focused. But when Montez went, I guess I just let go. When we were carrying him and trying to get him out, we felt such love for that man. I was sobbing. Keith finally had to get me moving. Keith was always in control. He had to be strong for the rest of us.

Payne was then forced to make another difficult decision, one that has haunted him since. His men were exhausted and, without knowing how far they had to travel or how close the enemy were behind them, he radioed for permission to leave Montez behind. Permission granted, they covered the body with branches and kept going. It was at least another hour before they staggered into the safety of the defensive position. They had travelled less than a kilometre but it had taken more than three hours.

The next afternoon, Payne was airlifted out with the wounded. He wanted Dellwo to go with him but there was no room. With the weather closing in, he would have to leave in the morning. Ever the conscientious medic, Dellwo told Payne: 'I'll check on you tomorrow.'

For Dellwo, tomorrow almost didn't come. That night the North Vietnamese bombarded the position with mortar fire. Dellwo was hit in the head, back and abdomen. One of his lungs was punctured and he nearly died before being evacuated to a Japanese hospital.

KEITH PAYNE RETURNED TO Brisbane a national hero. But the wounds of Kontum would never fully mend. Within ten years he was in the grip of alcohol and prescription drugs, his violent mood swings and brain snaps the result of undiagnosed Post Traumatic Stress Disorder. It was only the diagnosis and treatment of the condition in 1995 that allowed him to finally master his demons. He became an unofficial ambassador for the Victoria Cross and could speak matter-of-factly about the operation at Kontum. but he never shared his deepest feelings about the emotional toll the night had taken.

Dellwo went back to Spokane remarkably unscathed, except he couldn't speak Anastacio Montez's name without choking up.

And that is how it would have stayed, if not for a former Digger from Victoria. Lieutenant Colonel Jim Bourke had commanded 212 Company a year before Payne arrived in Vietnam. Like Payne, he had worked closely with Montez and considered him a friend. When he left the army, Bourke started Operation Aussies Home, an organisation that locates the remains of Australian soldiers missing in action. Bourke was instrumental in the location and repatriation of the remains of Flying Officer Michael Herbert and Pilot Officer Robert Carver, the last two Australians unaccounted for in Vietnam, in 2009. Eleven years earlier, he began researching Sergeant Montez's death, studying maps and interviewing survivors of the operation. In November 2001, he sent the results of his research to the Pentagon in Washington.

'I estimated his remains were 700 metres north of the officially recorded position,' Bourke says. 'I got a nice letter from the Pentagon saying thank you for your work and that they would look into it. I thought that was that.'

In October 2008, he took a phone call from JPAC in Hawaii. They had followed up his report. They had found something.

JPAC divides its work into four areas — Analysis and Investigation; Recovery; Identification; and Closure — each box needing to be ticked before work can begin on the next. Based on Bourke's report, in early 2008 members of the task force combed the area he had pinpointed and found what they term 'life support evidence', in this case a US Army issue metal buckle and some ammunition. It was enough to encourage them to return in the hope of getting what is officially called 'personal validation' — an eyewitness pointing to the spot and confirming that this was where the body had been left. In JPAC vernacular, this step is known as 'the smoking gun'. Payne and Dellwo were the guns.

In early 2010, their chance finally arrived.

On March 7, Payne left Sydney on a Vietnam Airlines flight. Dellwo was flown into Vietnam by the US military. They were reunited at Da Nang airport on March 9 and transported by road to the Hoang Van hotel in Kontum. For an entire day, they stayed at the hotel 'wargaming it'.

'We went through it over and over again,' Payne says. 'Every memory we had of that night, marrying it up with their maps and records. We didn't have much time out there. Everything had to be spot on.'

At 7.30am on March 10, Payne, Dellwo, Renas and a team that included five analysts, an anthropologist and a medic climbed into four-wheel-drive vehicles and took a trip into the past. The idea was to go to the defensive position and retrace their steps to the battlefield. Where once Payne and Dellwo had to be flown in and out by chopper, a road now leads right up to an area where the bunkers and shell holes are still visible.

Payne walked straight up to a bunker. 'Here's where we slept that night,' he called to Dellwo.

The younger man came over. 'Yeah,' he said. 'And right here is where I got hit.'

'It was pretty emotional,' Payne recalls.

'We didn't say much for a while after that.'

The two men had different memories of the direction they had come from. They split up, Dellwo going down the slope with Renas, Payne staying higher on the ridge with Renas' second-in-command, Sergeant David White. At the end of the first day, Dellwo accepted that Payne's recollection was correct.

'The thing was, I always knew the way we'd gone because I was the one navigating,' Payne says. 'My job was to get the men out of there. Jerry's was primarily to look after Monty and that's what made coming back harder for him. We both knew it would be emotional, but he had the memories of carrying Monty out and it was tough. I couldn't believe how much he remembered, but what he couldn't remember was the ground.'

With the vegetation so changed over the years, Payne steeled himself to imagine the land as it had been, not as it was. It forced him to retrace the night in graphic detail.

> I had to marry the ground to memories of what happened that night. There were things that I remembered that I hadn't thought of before. I'd pushed them to the back of my mind. Some of them were probably better off staying there.

By the end of the second day, Payne and Dellwo had agreed on the area where they'd left Montez, but after hours of what Payne termed 'rough slog' there was no sign of his remains. Importantly, they had established that the 'life support evidence' found in 2008 could not have been from Montez. 'It was too far away from where we were,' says Payne. 'And Monty didn't have any ammunition on him anyway.' To JPAC, that is vital information and, with Payne and Dellwo identifying a search site of 35 metres square, it could warrant another search.

'I don't think we'll start digging yet,' Renas tells me. 'It is very expensive work. More likely we'll have another investigation this time next year. It is a large area, but it's not impossible. I really feel for Keith and Jerry. Every now and then there were moments when you would catch it: in their minds' eye they had gone back to the past. They really want to bring this guy home.'

The mission might not have brought the result Payne and Dellwo hoped for, but they both left feeling a weight had been lifted. They knew they had done everything they could for Montez — just as they had on May 24, 1969.

'It did hurt having to leave him there,' Payne says. 'I asked myself again and again: *Was I right? Did I do everything I could have done?* The night after the map reccie, before we went out, I came to terms with the decision I made. All of the team that came out with me and Jerry, current American soldiers, said: "You made the right decision."

'It was a good settling experience. I can say now that I can rest easy. I can say I did the right thing without question. I always thought I did, but the question was always there. Now there's no question.'

Payne has been troubled over the years by dreams of Kontum, by flashbacks of what happened that night. Has the trip finally lifted the weight from his mind? Partly, he says. It will only be fully lifted when he receives a call from JPAC saying they have found Montez.

'I hope they don't take too long,' he says. 'I want to be young enough to go to his funeral.'

Chapter 6

THE GENERAL AND THE WOLF CUB

RON REES SITS ON a chair under a life-sized photograph of US General Douglas MacArthur and recalls the meeting that changed his life. It was January 1944, a Saturday. Six-year-old Ron and his older brother Norm, 12, were doing what they did every weekend during the war years — fishing for bream in the Brisbane River near Victoria Bridge. It wasn't for fun. With their father Hugh off serving with the RAAF, the two boys needed to supplement their family's meagre meat ration.

Around 5pm, Norm told Ron to pull in his line. Ron looked confused. It was still light and the fish were biting. They would usually stay until dark before catching the tram home to Rainworth, in the city's inner west. But this day was different.

'Come on Ronnie,' Norm said. 'We're going to see the general.'

General MacArthur, Supreme Commander of all Allied Forces in the South West Pacific Area, had arrived in Brisbane on July 20, 1942.

Just over four months earlier, ordered by US President Franklin Roosevelt to leave the Philippines — where his troops were engaged in fierce fighting with the Japanese — and set up his headquarters in Australia, MacArthur, his wife Jean, their four-year-old son Arthur and the boy's Chinese nanny Ah Cheu, and members of the general's staff had made a hazardous 720km voyage in four war-weary PT boats from Corregidor Island in Manila Bay to Mindanao Island.

From there they took off from a muddy plantation runway in two USAAF B-17 Flying Fortress aircraft dispatched from Townsville, North Queensland, for the 3200km flight to Darwin in the Northern Territory.

With Darwin under attack from a Japanese bombing raid as they approached, the two B-17s landed at Batchelor Airfield, 80km south. The general and his party were then flown on two commercial aircraft to Alice Springs. From there it was an 80-hour train trip, first to Adelaide and then Melbourne, with a water stop at Terowie, 220km north of Adelaide. It was there that MacArthur, who had been based in the Philippines since 1935, made his famous promise to waiting news reporters: 'I shall return.'

MacArthur, then 62, initially established his headquarters in Melbourne but after four months, with US and Australian troops marshalling in Queensland for the Pacific campaign, he elected to move it to Brisbane.

The arrival of MacArthur, his family and staff, as well as his Wolseley limousine with USA 1 numberplates, and desk, specially made in Melbourne, was a great day for the people of the Queensland capital.

There had long been rumours, never officially confirmed, that the government led by prime minister Robert Menzies until his resignation in August 1941 had formulated a plan to abandon the northern half of Australia in the event of invasion by the Japanese, and attempt to hold the southern half until help arrived from the US. The so-called 'Brisbane Line', stretching from Brisbane to Perth, was said to be the point of division.

Little wonder, then, that when the MacArthur family moved into the top floor of Lennons Hotel at the intersection of George, Ann and Adelaide streets, and the AMP building at 201 Edward Street was taken over for his headquarters, the residents of Brisbane breathed a mighty sigh of relief. Surely, they reasoned, the Americans would not establish their Pacific headquarters in Brisbane if they planned to hand the city over to the Japanese — a theory that was later backed up by MacArthur's stated policy that the enemy should be stopped in New Guinea.

It immediately made MacArthur the most popular man in Queensland: a warrior and a saviour. But he was more than just a soldier. He was a larger-than-life celebrity.

With his braided general's cap, swagger stick, sunglasses and corncob pipe, MacArthur was like a character from a Hollywood movie. It was an image he did nothing to dispel. A man noted for his machismo, self-confidence and ambition as much as for the military nous and courage under fire that saw him awarded the Medal of Honour, three Distinguished Service Crosses and seven Silver Stars, MacArthur lapped up the attention of the adoring Brisbane public.

He and Mrs MacArthur did not enter into the social life of wartime Brisbane — their only official outing during their time in the city was a luncheon at Government House hosted by the governor, Sir Leslie Orme Wilson, and Lady Wilson 10 days after their arrival — but the public craved a glimpse of their most distinguished visitor. While the local press rarely carried photographs of General MacArthur or his family for fear of alerting the enemy to his movements, his daily routine soon became common knowledge. He would leave Lennons each morning and be driven the short distance to the AMP building, arriving at 10am. He would work in his office until early afternoon, when he would return to the hotel for lunch and a short sleep before being driven back to the office at 2.30pm. He would then head back to Lennons at 5.45pm for dinner, before returning to the office around 7.30pm and working through until 10pm. Each time he walked to or from his car, under the watchful eyes of heavily armed US Military Police, a crowd gathered to applaud, offer greetings or simply get a glimpse of the man his Filipino staff had christened 'Sir Boss'.

On that Saturday evening in January 1944, among the crowd waiting outside Lennons for him to arrive for dinner were Ron and Norm Rees.

'I was holding a big bag of smelly fish so that cleared a path to the front of the crowd,' says Ron, now 81.

> Norm said, 'Now Ronnie, when he walks past you have to salute him because he's a general.' I didn't know how to salute, but I was going to be a wolf cub, and I knew the two-finger wolf cub salute, so Norm said, 'Alright, just do that then.'

At 6pm on the dot, the big Wolseley pulled up outside the hotel and out climbed MacArthur to the cheers of the crowd. As the general walked past him, Ron Rees came to attention and gave his two-finger salute.

> He looked down and smiled and said, 'Good boy, V for Victory.' And I said, 'No, mate. It's not V for Victory. I'm going to be a wolf cub. It's the wolf cub salute.'
>
> I was wearing an old school cap and I'd written 'US Army' on the front in indelible pencil. All the kids did that during the war. Norm had 'AIF' written on the front of his. He looked at my cap and said, 'What rank do you hold in the United States Army?'
>
> When I didn't answer, he reached into his pocket and pulled out a lieutenant's bar like they would wear on their collar. He handed it to me, saluted and said, 'You're now a lieutenant in the United States Army.'
>
> And I said, 'Thanks mate.'

When we got home, Norm dobbed me in. He told Mum I'd called the general 'mate' and I got a 20-minute lecture.

As he speaks, Ron is seated in the foyer of the MacArthur Museum, in the building now called MacArthur Chambers, which occupies the space that once housed the general's headquarters in the former AMP building. In his hand is a World War II US Army garrison cap. Pinned on the side is the silver lieutenant's bar handed to him by General MacArthur. Among his most cherished possessions, they symbolise 73 years of devotion to the memory of the wartime leader and the young men and women who served and died in the Pacific conflict.

The museum is busy the morning I visit. Last month's 75th anniversary of MacArthur's arrival in Brisbane has led to a spurt of interest in the general's life and times. It has been open since 2004, after then-Brisbane lord mayor Sallyanne Atkinson approved redevelopment of the office block into residential apartments on the proviso that MacArthur's former headquarters on the eighth floor be retained as a permanent memorial to the building's historical significance. It is administered by a trust and manned by volunteers, including Ron Rees. Artefacts on display include the control yoke from the aircraft in which Japanese navy commander-in-chief Admiral Yamamoto was shot down and killed over Bougainville in 1943.

There are replicas of MacArthur's cap and corncob pipe, and a medal of honour donated by former US president Bill Clinton. A wedding dress worn by an Australian 'US war bride' stands in a glass cabinet. But the most striking and emotive part of the museum is down a hallway and away from the main area — MacArthur's private office.

The former AMP conference room, it has been meticulously restored to how it looked when MacArthur sat behind his desk below a framed portrait of George Washington. While the original desk is now on display at the MacArthur Memorial Museum in Norfolk, Virginia, the one in Brisbane is an exact replica — 'right down to the cigar burns and coffee stains,' says the museum's executive director John Wright.

To stand in that room and look at the photographs on display of MacArthur with Commander-in-Chief of the Pacific fleet, Admiral Chester Nimitz, taken just a few steps away, is to feel a spine-tingling sense of history, but for a small group of people there was no need for photos or replica furniture. They had their memories.

Over the years, many of the Australian Army clerical and administrative staff who worked in the building during MacArthur's time visited the museum to finally get a peek at The Boss's inner sanctum. Ron Rees, who joined as a volunteer in 2005 after retiring from a career in the pharmaceutical industry, recorded their recollections.

One, who gave her name only as 'Carmel S', told of how MacArthur had noticed her, red-eyed from crying, as he passed her desk. A short time later, one of the general's aides came and asked what was wrong.

> I was 20 years old. I was to be married in three weeks. I had run out of coupons to buy a new dress for my wedding, and it looked like I would have to wear my uniform. I told him my story at 1.30pm and he told me to report to the front desk at 2pm.

When Carmel arrived at the desk, Mrs MacArthur was waiting.

> She told me she was taking me shopping. She took me over to [department store] Finney Isles, where the manager was there to greet us. I was in a state of shock, I thought I was dreaming. I kept telling them I didn't have any coupons.
>
> By 3.15 I had a dress to be married in, shoes, bag, gloves, plus a going away outfit.
>
> Mrs MacArthur said the general was taking care of the bill. I insisted on paying by instalments, but I was told, 'The Boss said it was a wedding present.'

Joan Williams and Val Parkin remembered getting into the lift on the ground floor of the AMP building just as MacArthur strode into the foyer.

> We got out but he called us back in. The army had gone into summer uniform but it was cold. The Boss asked us why we weren't wearing jackets. We told him it was against the rules. At 11.30am, a handsome young American soldier came to our desks with

> two packages. Inside each was a WAC jacket with a note: 'You can't be cold. Tell your CO I said so — MacArthur.'

And then there was the office messenger who jumped into the lift and bumped straight into MacArthur, who remarked that he must be late for a date.

'Yes, I am,' said the young man.

'Where and when?' asked the general.

When told that he was supposed to meet a nurse at southside Holland Park in 15 minutes, MacArthur said, 'Follow me.' He led the way to his car. 'Sergeant,' he said to his driver as they climbed into the back seat, 'would you drop me home, then drive this soldier to Holland Park.'

Dell Hicks was 10 years old when the MacArthurs arrived in Brisbane. She never met the general or saw the inside of 201 Edward Street until last year, but she became closer to his family than anyone else in the country. Dell's father, Sydney Ipsen, was caretaker of the Supreme Court building in George Street, across from Lennons. Her family lived in a cottage in the grounds. One day there was a knock on the door. Dell's mother answered it to find Mrs MacArthur on the other side.

'She had seen me playing in the yard from her apartment in the hotel and came over to ask my mother if Arthur could come and play.'

A play date was arranged and from then on Dell, then 11, Arthur, 5, and Neil Watt, 7, the son of Lennons' assistant manager, became regular playmates.

'I was at school but we would play every weekend,' Dell, 85, recalls.

> We would usually play around the courthouse, mostly running and chasings and ball games. One of the country judges, Judge Brennan, gave Arthur a goat and a gig. It was called Tojo, after the Japanese general who ordered the attack on Pearl Harbor. When the goat died, Neil and I used to pull Arthur around in the gig.
>
> Arthur was a lovely little boy. He wasn't a snob. He never went on about who his father was and it didn't mean anything to me. There was a war on, he was a general, that was it.

As well as playing with Arthur in her yard, Dell was a welcome visitor to the MacArthurs' residence at Lennons. At her Mt Gravatt home in Brisbane's south, she has photographs of her at Arthur's sixth birthday party and a Halloween party.

> There was always a sergeant at the lift to make sure no one else got in, but they would let me go straight up to their floor.
>
> They all had their own apartment. I remember Arthur had a lovely train set and a little piano or organ.
>
> He was very good at music and would have lessons every week. I used to go up there sometimes and we

would watch movies. They had a projector and the movies would come from America. That was very new to me. Everyone does it now, but back then if you wanted to see a film, you'd go to the movies.

One time there was a rodeo at the Showground and we went to that in the general's big car. It was just me, Arthur, Mrs McArthur and Ah Cheu. I used to have meals there sometimes and for dessert we'd have ice cream with chocolate sauce that Mrs MacArthur made herself, and I used to think that was absolutely wonderful.

When MacArthur left Brisbane in July 1944 to supervise land operations against the Japanese — fulfilling his promise to return to the Philippines on October 20 — his wife and son remained in Brisbane.

On January 26, 1945, MacArthur's 65th birthday, he established his headquarters at Tarlac on the island of Luzon. That day, in a letter on Lennons Hotel stationery, Mrs MacArthur wrote:

Dearest Sir Boss,
For your birthday I send all my love to you and may it help to form a mantle of protection for you. I love you more than you will ever know. May we be able to share in peace many more of your birthdays together.
God Bless you,
Jeannie

Her wish came true. On September 2, 1945, General MacArthur accepted the formal Japanese surrender aboard the battleship USS *Missouri*, ending World War II. Mrs MacArthur and Arthur left Brisbane to join the general — after saying goodbye to Dell.

'When the war ended they went to Japan,' Dell says. 'I never saw Arthur again, but Mrs MacArthur wrote me a letter saying that he had settled down at school but he missed me.'

She looks down at the letter and smiles. 'It was a simpler time then,' she says. 'You've brought it all back to me.'

For Ron Rees, the brief meeting outside Lennons wasn't the end of his relationship with General MacArthur — just the beginning. In 1961, by then an officer in the Australian Army Reserve, Rees travelled to New York.

> I had always wanted to see the Waldorf Astoria hotel, so I headed out there but I went to the wrong entrance. I expected to walk into a magnificent lobby, but I had gone to the entrance for the permanent residents.
>
> I was thinking, this isn't too flash, when the lift door opened and out stepped General MacArthur. I don't know why I did it, it must have been something ingrained in my memory, but I stood to attention and gave the two-fingered salute.
>
> He looked at me and said, 'Well, I'll be damned, it's the wolf boy from Brisbane.'

MacArthur asked Rees if he had time for a quick coffee and led the way to the Waldorf Astoria cafe. The 'quick coffee' lasted two hours as the two men forged a bond.

When Rees returned to Brisbane, he wrote to MacArthur. The general's reply letter, dated April 16, 1962, which Rees keeps in a bank deposit box, is cherished on par with the lieutenant bars and the desk that MacArthur used at Lennons, which Rees obtained after the hotel was demolished in 1973.

Two years before his death at the age of 84, MacArthur wrote:

> My friend, I shall now close. I want you to know my old battered cap sits on a shelf beside my desk. Each time I look at it, many memories spring to mind.
>
> One of the fondest is of the little boy from Brisbane in 1944 who saluted me and called me 'mate'.

Chapter 7

SEARCH FOR A LONELY SOLDIER

I ASK HER IF it was hot when she found Charlie. 'Have you ever been to Charters Towers?' she says. Yeah. Stupid question.

It was in the high thirties and she'd been out in the paddock for hours. She only had one bottle of water and a stick she'd picked up off the ground to dig with. That, and a few house bricks and a couple of little Australian flags she'd thrown in the back of the car — just in case. All she had to go on was a list of names and numbers, and a rough idea of where he might be, which turned out to be wrong anyway.

'I wasn't looking in the right place at first,' she says.

'I'd been there half a day, at least. I was digging away in the dirt, my stick was breaking. It was bloody hot. There was only one tree, and that was a long way off …'

Finally she pulled away a few tufts of grass and scraped away the dry red soil on the top of a stone plaque, just as she had scraped dozens of other stone plaques, many broken and chipped by the slashers that come through every few months. One by one the numbers became clearer. A one, a three and a six, then a two, and finally an eight … she finished brushing the stone off with her hand, looked down at her list, then back to the dusty plaque. It was like checking Lotto numbers. She'd won the jackpot.

There below her feet lay the remains of Lance Corporal Charles Tednee Blackman, serial number 2584A, B Company, 9th Battalion, 3rd Brigade, 1st Division, AIF. Buried and forgotten for 50 years. She stood there for a while, not really knowing what to do next, then went back to the car and got the bricks and the flags. She put the bricks around the grave, then placed a plastic flowerpot on top and stood the flags upright inside it. Then she sat down for a chat.

'Oh, Charlie,' Suzanne said, the tears starting to flow. 'I've been looking for you a long time.'

Suzanne Murdoch, her daughter Carmen, sister Kaye and brother-in-law Chris Bull weren't the only people searching for Charlie Blackman. Far from it. The story of the three Blackman brothers, Charles, Alfred and Thomas, all of whom fought in World War I, has fascinated researchers since the surge of interest in the contribution made by Indigenous soldiers in the Great War began in the 1970s. In fact, as 'half-caste' Aborigines, and prodigious letter-writers, the Blackmans have for years been something of a 'go-to' family for historians seeking examples of Indigenous life and military service in the early 1900s.

Removed from their mother as children in 1902 under the Aboriginal Protection Act and placed in the settlement of Barambah — later renamed Cherbourg — 250km northwest of Brisbane, the three brothers volunteered to serve in the AIF in World War I. Alfred, the eldest, was killed during the Battle of Passchendaele in October 1917. Charles and Thomas survived and at war's end returned to Australia believing their military service would improve their status in the eyes of the authorities. Instead, they both ended their days as 'inmates': Thomas back at Cherbourg and Charles in an asylum at Charters Towers.

'They are case studies,' says historian Philippa Scarlett, author of *Aboriginal and Torres Strait Islander Volunteers for the AIF: The Indigenous Response to World War One.*

'They encompass all elements of the Aboriginal experience of that time.'

Scarlett, who has referred to the Blackman brothers in several of her published articles, is one of numerous researchers who have delved into some aspect of their lives, each adding a link in the chain that led all the way to that dusty paddock at Charters Towers. All have had their own motivation; most have never met each other, but all for their own purposes have helped uncover the story of this ordinary yet remarkable family.

Merv Hopton, 67, a semi-retired businessman from the Bundaberg suburb of Buxton, dabbles in historical research 'to keep my mind active'. He first came across references to the Blackman family when researching pastoral 'runs' on properties around Bundaberg. His investigations brought him into contact with Graham Jennings, a relative of the Blackmans who has carried out his own research on the family tree through oral history. Together they ascertained that Thomas Blackman Senior, the father of the three brothers, owned 100 acres (40 hectares) at Tiaro, a town on the Bruce Highway, 30km south of Maryborough and 130km from Hopton's home in Bundaberg. Son of a white father and Aboriginal mother, he died from cancer in 1898 at the age of 48, leaving no will. Within a few years his wife Emily and their children were destitute and living in a derelict house in Bundaberg.

'I'm of the opinion she got ripped off,' says Hopton, who has located a wealth of official documents relating to the family, including the report of Constable James Manning, who was sent to check on Emily and the children on September 16, 1902.

'The above mentioned is residing in an old dilapidated building in Grafton St with her six children aged from 14 to 3 years,' it reads. 'This woman has been a resident of Bundaberg for about 12 months and has been trying to make a living for her children and self as a charwoman. She is in very poor circumstances. The house is almost devoid of furniture and bears the aspect of extreme poverty. The constable is of the opinion that this woman is not a fit and proper person to have charge of these children.'

Two days later, an application was approved at Bundaberg Police Court for 'disposal' of the children — listed as Louisa, 14, Alfred, 11, Thomas, 10, Mary, 8, Charles, 5, and their half-brother Willie, 3 — to Barambah. Established in 1900 on a former cattle station 250km northwest of Brisbane, Barambah was a government-run settlement at which Aboriginal children were separated from their parents and older relatives. The adults lived in camps away from the main administration buildings while the children, housed in dormitories, were taught to speak, read and write English in preparation for life as domestic servants or farm labourers for white property owners. As part of the 'inmates'' assimilation with white culture, Barambah entered cricket and rugby league teams in district competitions. The settlement produced many outstanding athletes, including rugby league star Frank 'Bigshot' Fisher — grandfather of Olympic champion Cathy Freeman — and fast bowler Eddie Gilbert, who dismissed Don Bradman for a duck in 1931. It also provided 47 young Indigenous men who joined up to fight in World War I. Their stories were told in an exhibition titled 'The Boys From Barambah' at Cherbourg's Ration Shed Museum.

The first soldier from Barambah to enter the war, in fact one of the first Aborigines to enlist anywhere, was 19-year-old Charles Blackman, who joined up at Childers on August 18, 1915. At the time Charles was working as a labourer at Illoura, the property of Mr John 'Jack' Salter at Biggenden, 340km northwest of Brisbane. While there, Charles became very close to Mr and Mrs Salter and their three children, William, Charles and Margaret. During his time in uniform, Charles wrote regularly to Mr Salter. The correspondence was kept by the family and in 2001 donated to the Australian War Memorial in Canberra by the Salters' grandchildren. Now part of the Military History collection, the letters and postcards provide a fascinating and at times chilling insight into the life of an Indigenous soldier in World War I.

The first letter, addressed to 'Dear friend Mr Salter', was sent from Fraser's Paddock, Enoggera, in Brisbane's northwest, where Charles was training with the 6th Reinforcements of the 25th Battalion.

> Just a few lines to let you know how I am. I am doing all right. Sorry to say that I was very sick from the inoculation. I was off my feet for a few days. A fine young fellow died the other day from the same thing. The 9th Battalion left us last week. They left for Sydney. That's where we will be going soon. There are a few thousand here now and there are a lot of different kinds of fun here. How are they all down that way? Are they all well? I hope so. How are the children? Are they all right, and how is Mrs Salter? I hope she is well. We had a very big day on the Exhibition Day in Brisbane. We were watching

> all the morning. The crowd in the streets stretched nearly three miles. We couldn't get around the Show for crowds and there were three bands playing. I think that is all to say this time. God bless you all. I'll try and come and see you before I go off. Goodbye till we meet again by and by.

A postcard to the children from 'somewhere in France' follows in the New Year, before a letter that hints at the horrors of war that Charlie has experienced. In March 1916 he was transferred to B Company of the 9th Battalion, a battle-ravaged unit that had been the first force ashore during the Anzac landing at Gallipoli almost a year earlier. Days after he joined the battalion it headed to France and the Western Front. One of the first major confrontations with the Germans was in July at the Battle of Pozieres in the Somme. The Australian War Memorial notes that 'few came out of Pozieres without physical or mental scars'.

To the Salters, Charlie wrote:

> Dear friends, I received your nice parcel which I thank you very much for. The scarf is very warm and acceptable. I only wish I could parcel myself home now. I have been very lucky according to what I have been through. Pozieres was terrible but I'll return.

From Pozieres, the 41st Battalion fought at Ypres before being entrenched opposite the Hindenburg Line, the

140km concrete, steel and barbed wire defensive line the Germans built along the Western Front from Arras to Laffaux. By the end of the next year, Charlie had been to England on leave ('I had a glorious time there — people in England think the world of you and they take you all over the place and show you everything'), made a good mate named Frank and narrowly avoided capture.

> Frank and I was sleeping in the same dugout cooped up like sardines and the Prussian guard made a big attack and when they came over, what do you think happened? Well, the first dugout they came to was the one me and Frank was sleeping in but we were relieved about an hour before they came over. We went back to supports and if we weren't relieved we would have been prisoners of war in Germany. They took a few of our fellows once but they made a miraculous escape and got back to us. This is not half of what I could tell you about the Battle in France. What a soldier don't know is not worth knowing. We have seen some sights of all kinds. So far we haven't seen any better places than Australia. France is beautiful but it rains too much. You never see the sun here. One thing here is that clouds never need to thunder. The guns do all the thundering and the flashes they make is just like lightning. They make more noise than the thunderstorms in Australia. I would rather be in a thunderstorm than a battle.

What Charlie didn't know when he wrote that letter on October 29, 1917, was that his oldest brother Alfred had been killed three weeks earlier in the Third Battle of Ypres, a campaign of skirmishes and battles that lasted four and a half months and included the Battle of Passchendaele. It cost an estimated 40,000 Australian casualties.

The two older Blackman brothers were working on properties around Childers, 325km north of Brisbane, Alfred as a labourer and Thomas a stockman, when they joined up in the summer of 1916–17. Like Charlie, they were well thought of in the district and on January 26 a reception was held at the Queens Hotel to 'bid them farewell', as the *Maryborough Chronicle, Wide Bay & Burnett Advertiser* reported.

> Mr Monty Parker occupied the chair, and there was a fair attendance. The chairman pointed out that there were three brothers — Charlie Blackman, who had volunteered a year ago, and now the other two brothers were going — and that the male members of the family would be doing their bit. He called on Mr Whitford to make a presentation, which that gentleman did, pointing out that another two stalwart natives of the district were going to the front animated with a desire to be found fighting for the flag. He trusted they would have the opportunity to distinguish themselves and maintain the Anzac fame and that we could have the pleasure of welcoming them home again on a safe return.

> The presentations were a tobacco pouch and Meerschaum pipe, in case, to each. Mr Stirling, in proposing the toast to the health of the young men, said he hoped the exigencies of warfare would make the brotherhood of Australians all the stronger and better animated with one spirit and one aim. There was no doubt victory was ahead, but strenuous work was needed as the enemy had been long preparing and was themselves drilled and armed. It would only be by the execution of greater power that the end would be attained. The toast was duly honoured and each of the brothers acknowledged the compliment, thanking the donors for the presents and promising to do their bit when the time came. If they came home again they hoped to meet in Childers.

Alfred and Thomas Blackman joined the 41st Battalion 7th Reinforcements and sailed out of Sydney Harbour on February 7, 1917, aboard HMAT *Wiltshire*. By July, they had joined the main force of the battalion at Flanders and at dawn on October 4, 1917, were part of the British force preparing to attack German lines at Broodseinde Ridge, in what was the first engagement of the Battle of Passchendaele. The Australian troops were heavily shelled even before the order to attack was given, and nearly 1000 were wounded. Alfred Blackman was one of them, suffering shrapnel wounds to his neck. He was treated in the field by members of the 3rd Australian Field Ambulance and evacuated to the 7th Canadian General

Hospital at Etaples, 27km south of Boulogne. He died there four days later and is buried at the Etaples Military Cemetery. He was 27 years old.

When the Blackman brothers' mother Emily died in Maryborough in 1928, aged 56, among her few belongings were Alfred's World War I victory medal, and a booklet about the Etaples Military Cemetery containing a photograph of his headstone, for which she had paid sixpence.

By the time Alfred and Thomas joined up, their younger brother Charles was already a hardened veteran. On November 23, 1917, a week after being promoted to Lance Corporal, he wrote to Jack Salter during a spell away from the fighting:

> The last time up in the line I killed five Germans. I did it because they held up their hands till we got about 10 yards off them and then the dirty brutes threw bombs at us. So that's why we killed them. If they hadn't thrown bombs at us we would have taken them prisoners and they would have been all right. This totals 10 for me since I've been in France. That's a good haul for one man isn't it? I have done my bit, I think. The next time up I'll get some more and I'll kill all I can.

Over the next 10 months Charlie continued to write to the Salters about his experiences, about friends made and lost, and how lucky he was to still be alive.

> Last time in we had some stiff fighting. My cobbers are away wounded. I were nearly going away wounded too but I am wonderfully lucky. I wish other soldiers were as lucky as me.

In August 1918, Charlie's luck ran out. During the Battle of Amiens, the start of the so-called Hundred Day Offensive that led to the end of the war, he was exposed to chemical gas and repatriated to England. As part of treatment that would continue for months, he convalesced at Greenhill House at Sutton Veny in Wiltshire. The stately home, donated to the war effort by the Walker Whisky family, was linked to the Sutton Veny Military Hospital, where during or immediately after the war 141 Australian soldiers and two nurses died either from wounds or the influenza pandemic. To this day, children from Sutton Veny Primary School celebrate Anzac Day in honour of the 143 Australians buried in the church graveyard.

During his time at Sutton Veny and then back in France and Belgium, Charlie continued to write to the Salters, the self-described 'lonely soldier' grateful for their letters and counting the days until his return:

> You are the only person I get letters from. Mother never writes to me, or any of the others. I'm glad I've got someone like a Father and Mother to me. I wish I was returning to Australia today.
>
> I think this war is a nuisance. It's keeping us away too long. I haven't seen any place like Aussie in this

> world. We are all hoping to get back again. Every day you can hear dozens of fellows saying, I wish I was in Aussie now. The reply is, you've got a lot of mates, Digger.

Charlie's wish came true when he returned to Australia on June 5, 1919. He was medically discharged due to the effects of his gassing on July 30. Thomas had returned and been discharged four months earlier.

For the two surviving Blackman brothers, civilian life would be tough, although as Merv Hopton says, 'the same could be said for a lot of men who returned from the war, white and black'. Away from the structure of army life, where every move a soldier makes is documented and recorded for posterity, it is not easy to follow the path Charlie's life took, but Hopton has spent a great deal of time and money following a paper trail that, together with documents held by National Archives and the Australian War Memorial, gives a broad picture of Charlie's later years.

It is believed Charlie returned to Biggenden after the war and resumed his work at Illoura. An undated note in the Salter collection at the Australian War Memorial suggests he moved on without notice: 'Don't search for me because I have left for some distance away. I will leave a word in Biggenden soon as I can.'

He is next heard of living around Tully, 200km north of Townsville. The Australian War Memorial lists him as leaving Biggenden in the mid-1920s and working on a property at Feluga, near Tully, and there are records of him working on a council gang, building and repairing timber bridges. In February 1936, he wrote to the Officer-in-Charge, Victoria Barracks, Brisbane, requesting a copy of his discharge papers: 'Please could you kindly let me have a copy of my discharge as I have lost mine and since people who give us preference on jobs usually ask for them I would be glad to have a copy of mine sent to me as soon as you can send it as delay might cost me a good job.' In a statutory declaration he said the discharge document had been lost 'while in the shifting of camp from one part of the bush to another during the time I was employed by the Cardwell Shire Council'. Eight months later, he took a lease on a small plot of government land at Tully, as was a common practice for cane cutters working in the area at the time.

In October 1943, a public notice was placed in *The Cairns Post* newspaper advising that Charles Tednee Blackman intended to change his name by deed poll to Charles Thomas Graham, Thomas being the name of his father and Graham his mother's Emily's maiden name. The reason for Charlie's name change remains a mystery.

'No one seems to know,' says Hopton. 'He wasn't hiding from anything or anyone. There is no mention of his name in any police records. He was clean in that regard. Emily Blackman raised three fine young men. She did everything she could to keep her family together, and that's the tragedy of it.

'All we know about Charlie's last days is that he ended up in Mossman Hall mental facility at Charters Towers.

What triggered that we don't know, but the feeling is the war was to blame. He suffered severely from shell shock.'

Thomas Blackman also had an unhappy end. In 2009, a box of long-lost documents belonging to anthropologist Caroline Tennant-Kelly, who had died in country NSW 20 years earlier, was discovered in the shed of an old friend. In the 1930s, Tennant-Kelly had studied the living conditions of Aborigines housed at Cherbourg. Among the notes and photographs found from that period was a letter written to her by Thomas Blackman on February 22, 1935:

> I'm writing to you referring to the treatment of half-caste soldiers. I'm a half-caste soldier myself. There were three of us went to the Great War out of my family. One was killed at war and two of us returned after the war was over. I always thought fighting for our King and country would make me a naturalised British subject and a man with freedom in the country but I have hardly had freedom since I returned from the war. I have no justice at all. I don't know if the King knows how his half-caste returned soldiers are treated in this country. I was brought to Cherbourg May 7th, 1934, and still find myself here. Is this the sort of freedom that I went to France and helped fight for? I say no. I want justice. I'm a man that is well liked wherever I worked. I was living at Maryborough and had my good home broken up by the Chief Protector of Aboriginals. It hurts me very much.

> I was locked up for seven days for being intoxicated and using bad language but it seems the Chief Protector wasn't satisfied with me taking my punishment but places me under the Act and put me to a settlement like a dog. I was brought to this settlement and find the Chief Protector hasn't got any accommodation for my wife. Another thing I find is that a half-caste soldier has got to get a permit from the Superintendent to go about the country. We helped to fight, same as white soldiers did. Think of all the half-caste soldiers that were killed at war. What thanks have the half-caste soldiers got for going to war? We were good men at war but looked down on now the war is over.

Thomas is believed to have remained 'under protection' at Cherbourg until his death.

In early 2015, Merv Hopton obtained a copy of Charlie Blackman's death certificate, stating he had died at Mossman Hall Special Hospital, Charters Towers, from the effect of pneumonia on August 12, 1966. He was 65 years old and single, and had been unwell for four years. He was buried in the open grassed section of Charters Towers Cemetery.

In April 2015, Hopton visited Charters Towers library and spoke to historian Michael Brumby, telling him of the work he had done researching Charlie Blackman and that he was buried in an unmarked grave at Charters Towers Cemetery. Brumby, 63, contributes a monthly historical column in the local newspaper, *The Northern Miner*, and wrote his next column about the Indigenous World War I veteran buried somewhere at Charters Towers.

And that is where the final link in the chain that led to Charlie Blackman began to be forged. 'Suzanne read the story and she sent it to me,' says Kaye Bull, 61, who, along with her husband Chris, 71, an ex-career soldier, is a welfare officer at Airlie Beach RSL. 'Maybe as an ex-nurse it's the nurturing gene in me, but I just thought, this isn't right. This fella fought for his country and he deserves a proper burial. So we starting searching.'

Looking for Charlie became a family affair. Suzanne's daughter in Sydney, Carmen Schegg, 44, got to work on *ancestry.com*. Kaye and Chris went through military records. While they were following a path well worn by the likes of Philippa Scarlett, Merv Hopton, Graham Jennings and Eric Law from the Boys From Barambah project, they got one step further than anyone else, including finding Charlie's great niece, Sharon Blackman, in Ayr, in North Queensland.

The breakthrough came when they learnt of the name change and tracked down a grave number for Charles Graham, rather than Charles Blackman. Armed with the list of numbers sent to her by Kaye, and one bottle of water, a stick and some plastic flags, Suzanne, 65, finally found the lost Digger on July 15, 2016.

Kaye and Chris have received funding from the RSL to place a headstone over Charlie's grave and consecrate it at a ceremony attended by relatives, Cherbourg residents and representatives of the military. It was a promise made by Suzanne as she sat talking to him that hot July day.

> I said to him, 'I know who you are, Charlie. I know what you did and we are going to make sure you get the respect you so richly deserve. Every night at the RSL clubs they say, "Lest we forget." Well, you were forgotten, Charlie Blackman, but we're going to make sure you're not forgotten any more.'

***

Postscript

A white marble gravestone was dedicated with full military ceremony over the burial site of Charles Tednee Blackman at Charters Towers on Thursday, October 12, 2017. Among those in attendance were members of the Blackman family and representatives of the Australian Defence Forces, including Major General Mark Kelly, former commander of all Australian forces in the Middle East and Afghanistan.

Speaking at the graveside, Kaye Bull, who along with her husband Chris was a prime mover in the search for Lance Corporal Blackman's last resting place, said:

> We go to every Remembrance Day, we go to every Anzac Day and say, 'Lest we forget.' But we do forget

some of these people. We have to recognise them. It is as simple as that. They put their lives on the line for us and it doesn't matter what colour you are. At the very least, anyone who puts on a uniform deserves a proper burial.

Chapter 8

AUSTRALIA'S PEARL HARBOR

MARTIN PASH REMEMBERS IT as if it were last night. He remembers the ship shuddering, the smoke and flames. He remembers heading for the gangway and a fellow merchant seaman named Rayner calling out to the men to grab their life vests: 'Don't forget your Mae Wests, boys.'

He never saw Rayner again.

He remembers trying in vain to release the lifeboats, then being sucked back down the hatch as the ship went under. He remembers seeing a giant illuminated red cross and desperately pulling himself towards it. He remembers 36 hours on a raft and the deathly silhouette of a Japanese submarine in the half darkness. Most of all, in his waking hours and in the dreams that have plagued him for decades, he remembers the screams.

It was May 14, 1943. Pash, then 20, was a steward on the hospital ship AHS *Centaur*, which was heading up the coast from Sydney to New Guinea with medical staff and 149 men of the 2/12th Field Ambulance on board.

At 4.10am the ship, believed to have been about 24 nautical miles northeast of Stradbroke Island's Point Lookout, was hit by a torpedo from the Japanese submarine I-177. Most of the 332 people on board were asleep. It took the *Centaur* less than three minutes to sink, and in that time fewer than 100 escaped. Just 64 survived their wounds, sharks and almost two days adrift before being rescued by the destroyer USS *Mugford* and brought to Brisbane.

Of the 18 doctors aboard, Lieutenant Colonel Leslie Outridge of Gympie was the only survivor. Only one of the 12 nurses, Sister Ellen Savage, survived. The dead included three brothers.

The *Centaur* has not been sighted since. But its sinking has been a continuing source of heartache, rumour, mystery and a painful case of mistaken identity. Now, thanks to the efforts of the descendants whose lives were changed forever on that morning in 1943, it is the target of a $4 million search by the world's foremost shipwreck hunter.

Over the past 14 years David Mearns, a New Jersey-born marine scientist based in Britain, has found 21 major shipwrecks. For Australians, by far the most important of these was the HMAS *Sydney*, the light cruiser sunk by the stricken German auxiliary cruiser *Kormoran* off the West

Australian coast in November 1941 with the loss of all 645 aboard.

It was just days after locating the *Sydney* in 2008 that Mearns, 50, received an email imploring him to find the *Centaur*. In December, funded by $2 million each from the federal and Queensland governments, he began the detective work he believed would lead to the ship's final resting place.

THE SINKING OF THE *Centaur* was Australia's Pearl Harbor, our World War II version of 9/11. That it happened to a clearly marked and well-lit hospital ship off the Queensland coast brought home the closeness of the war and the ruthlessness of the enemy.

The disaster became a rallying cry, a call to arms. To many it was the thought of the nurses who were sleeping midship, close to where the torpedo exploded, that was hardest to bear. A poster depicting Sister Savage clinging to wreckage as the *Centaur* went down exhorted Australians to *'Work, Save, Fight and so Avenge the Nurses!'* As the years passed and Australians got on with their lives in the prosperity of peacetime, memories of the *Centaur* faded for all but a small group — those like Martin Pash who had survived the sinking, and those whose loved ones had not.

Keith Clegg was seven years old when his father sailed away on the *Centaur*. Percy Clegg, 39, was a farmer from outside Warwick who joined up in 1942 and trained as a theatre technician. An only child, Keith and his mother

Priscilla moved into a house in Warwick to await Percy's return. He never came back.

Official word of his death came via telegram. 'I vividly remember being at school and the headmaster walking into the classroom with a grim look on his face,' Keith Clegg recalls. 'He told me to go home but he didn't say why. I knew it was something bad and I walked very, very slowly, hoping that if I didn't get there it might all go away. And then I got home ...'

At that moment he and his mother became what he calls a '*Centaur* family'.

> The *Centaur* haunted my mother all her life. Because the search didn't happen for 36 hours she always wondered: *Did he get out, was he in the water?* She thought maybe, if they'd just got there earlier, he might have been saved. She continually tried to talk to survivors, trying to find out if anyone had seen him, or knew what happened to him. She died in 1979 aged 73 still wondering. She never had closure.

Over the years there have been several memorials to the victims, and annual services held on the anniversary of the sinking — the largest at Sydney's Concord Hospital and Melbourne's Heidelberg Repatriation Hospital, where many of the ship's nurses and doctors had trained. In Brisbane, the late Vince McCosker, a survivor, organised an annual service and worked selflessly to raise funds for a permanent memorial, a dream that was realised in 1968 at Wickham Point, Caloundra, with the help of the local Rotary club.

In 1993, another memorial was unveiled at Point Danger, close to the Queensland-NSW border. The nearby public school was named Centaur Primary and each year the students organise and officiate at a memorial service.

It was at the inaugural Point Danger service that Jan Thomas first met others affected by the sinking of the *Centaur*. Thomas' father, Captain Bernie Hindmarsh, had been the GP at the NSW mid-north coastal town of Macksville. The town 'lent' him to the war effort and was devastated by his death.

'I'm one of the lucky ones,' says Thomas. 'I was six when my father died, so I have cameo memories. My younger sister has slight memories and my brother was 18 months old. He can't remember anything, but my father was so prominent in the town and so many people have memories of him that they share with us, even today. A lot of others who lost family members don't have that.'

At that first Point Danger memorial service, the families shared experiences and gained strength from each other. From there they formed the *Centaur* Association. Conceived as a self-help group to nurture the healing process, it was also the first step towards looking for the shipwreck with a view to having it officially declared a war grave.

The Mearns expedition is not the first to search for the wreck. In 1995, diver Don Dennis announced he had found it in 170 metres of water, 17km off the lighthouse on Moreton Island.

In 2002 it was discovered that the wreckage was actually that of the *Kyogle*, a freighter used by the Australian Air Force for target practice in 1951.

'A lot of people were hurt by that,' says Thomas. 'They had laid wreaths and scattered ashes over the site. It made some of us in the association more determined to find the wreck, but until the *Sydney* was found it was just a pipe dream.'

Ian Hudson was the godson of Lieutenant Colonel Clement Manson, commanding officer of the hospital corps attached to the *Centaur*. Manson was last seen by Sister Savage and fellow nurse Merle Morton just before the ship sank.

'In that instant the ship was in flames,' Sister Savage told reporters from her hospital bed. 'We ran into Colonel Manson, our commanding officer, in full dress even to his cap and Mae West lifejacket, who kindly said, "That's right, girlies, jump for it now."

'The first words I spoke were to ask, "Will I have time to go back for my greatcoat?" as we were only in our pyjamas. He said no and with that climbed the deck and jumped and I followed ... the ship was commencing to go down.'

Hudson was two and a half years old when Manson died, but his father, Charles 'Alec' Hudson, spoke of his lost friend often.

'My old man and Clem Manson were great mates,' he says. 'He and my mother would tell me stories about him. I never did anything about it when I was growing up but when I got older, I suppose I got pangs. I wanted to get involved.'

In 2003, Hudson drove 1100km from his home in Airlie Beach to the Caloundra anniversary service organised by Erica Costigan and the Caloundra RSL *Centaur* Committee. Afterwards, he joined the *Centaur* Association and five years later he was the one who sent an email to Prime Minister Kevin Rudd calling on the government to fund a search for the ship.

'I'm a mechanical engineer,' he says. 'When David Mearns found the *Sydney* I realised that the technology was available to find the *Centaur*. We contacted David and when he replied that he was interested I wrote to the PM. The association didn't have any money; we were really just dreaming but maybe because I'm a bit younger than some of the members and have a bit more testosterone, I thought I'd give it a go.'

The prime minister was contacted on April 10, 2008. Just over two weeks later, Hudson's email was forwarded to the office of Queensland Premier Anna Bligh. On May 14, 2008, the 65th anniversary of the sinking, Bligh made a statement in Parliament acknowledging the sinking of the *Centaur* and wrote to the prime minister proposing a joint initiative to locate the wreck. Following a cost analysis, a joint statement was released announcing that $4 million in funding had been approved. The date, coincidentally, was December 7, 2008 — 67 years to the day since the Japanese bombing of Pearl Harbor.

'The first step is to go to the original primary source documents,' says wreck-hunter Mearns, who beat ten other tenderers for the job. 'From there I go to oral testimony and see if it stands up to scrutiny.'

His primary sources are the final bearing taken by second officer Gordon Rippon before the ship was torpedoed, and evidence given to a War Crimes Commission inquiry in Melbourne in 1944 by ship pilot Richard Salt, who swore he saw 'the glare of the Cape Moreton light ... over 27 miles from the land' as he floated in a life raft.

Mearns has gone through the testimony of other survivors and experts 'with a fine-toothed comb'.

He interviewed 86-year-old Martin Pash — one of three remaining survivors but the only one with cogent memories of the incident — in Melbourne and flew to Montreal to speak with Canadian academic Christopher Milligan, whose uncle, David Ireland Milligan, perished on the *Centaur.* Dr Milligan spent 17 years researching the sinking, and together with Queensland marine historian John Foley published the book *Australian Hospital Ship Centaur: The Myth of Immunity* in 1993. Mearns plans to travel to Japan in the hope of finding official reports and even the logbooks of submarine I-177.

What he finds in Japan will be of great interest to historians who have speculated as to why submarine commander Hajime Nakagawa would order an attack on a brightly lit hospital ship.

The War Crimes Commission was told by survivor Salt that before the *Centaur* sailed from Sydney, union delegates from the crew demanded an audience with the captain to discuss 'rumours that had been going round the waterfront that the ship had personnel or cargo of arms'. They were told by the ship's master Captain Murray that 'to his knowledge, every condition of the Geneva Convention had been complied with'.

The commission heard that the only arms carried on the ship were 50 rifles and ammunition issued to drivers in the 2/12th Field Ambulance. The rumours, however, persist to this day. There was talk that the *Centaur* was smuggling troops disguised as field ambulance personnel to New Guinea; that because it had brought several wounded Japanese back to Australia on its previous voyage from New Guinea — perhaps to be interrogated — it had relinquished any right to immunity.

Another theory, the one held by Pash, is that the ambulance personnel standing on deck in their khaki uniforms as the ship left Darling Harbour might have given spies the wrong impression.

'We sailed down Sydney Harbour like a troop ship,' Pash says. 'What would a spy have made of that?'

Commander Nakagawa was jailed for four years as a war criminal for ordering his men to machine-gun survivors of three British merchant ships sunk by the I-177 in February 1944, but he was never charged over the sinking of the *Centaur*. He died in 1986 without ever speaking publicly about the incident. The silence of Japanese authorities, who didn't identify I-177 as the submarine responsible until 1979, only fuelled the rumours.

Mearns doesn't want to know why or how the *Centaur* went down, just where.

> Why a Japanese submarine wanted to attack what was obviously a hospital ship is of great interest to historians, but my only focus is on any navigational clues that pinpoint where the ship is. The only bona fide position I have to really go by is from second officer Gordon Rippon. He was backed up by Salt, the pilot who was on his last trip.
>
> These were two very experienced navigators with 40 years between them, giving not only context but navigational information such as lights, course and speed. Anyone saying differently has got to have pretty strong proof to override an officer on the ship.

While the testimony of Rippon and Salt has given Mearns a broad corridor in which to search, there are other clues.

> We know where the *Mugford* picked up the lifeboats, what time they picked them up and the time the *Centaur* sank. In a four-part equation, that's three we know but the one we don't know is where the ship sank. I can calculate that from the drift speed and direction that the rafts carrying the survivors travelled.

> What we have to do is re-create the current direction and speed and from that corroborate the area. It is an exercise I have done in the past, but we perfected it in the hunt for the *Sydney*. It was instrumental in us finding the *Kormoran*.

Mearns will also work with the weather experts who carried out the calculations that led to the *Sydney*, with the Australian Bureau of Meteorology planning a wind 'hindcast' test to re-create the conditions of May 14, 1943.

What are their chances of locating the *Centaur*? 'We've got a good shot,' says Mearns. 'It is findable.'

> There is no guarantee, but it is definitely worth trying. The most important aspect of finding a shipwreck such as this is to provide comfort to people who have lost loved ones. Locating the wreck goes a long way to helping them accept their loss. They know the wreck is there, they see a picture and finally everything can be put to bed. They can find some inner peace.

FOR SOME, THERE WILL never be closure. 'The sinking of the *Centaur* changed my mother completely,' says Jan Thomas.

> Before that she had been one of the leaders of the town. She was into everything, the golf club ...

> everything. When the *Centaur* went down, that all changed. She cut herself off.
>
> She wasn't like some people who were interested in when and where and what happened. She was interested in why. She was torn by all the rumours. She wrote letters to anyone and everyone who might know the truth. She wanted to put those rumours to rest, and she never could.

Sitting in the loungeroom of his neat home unit in the northern Melbourne suburb of Ivanhoe, Martin Pash is matter-of-fact when recounting the facts of that night 66 years ago.

'To me she's just another ship,' he says.

It is later, on a visit to nearby Heidelberg Repatriation Hospital, that the facade lifts and the memories prove too much. Pash shows off the Anzac Memorial chapel with the plaque he unveiled in memory of his shipmates on May 20, 2001, and the tribute to Alice O'Donnell, Ellen Rutherford and Jennie Walker, the nurses from the hospital who were lost on the *Centaur*. He leads the way down the long corridor of the *Centaur* Wing, his gait rolling like the old seaman he is, until he comes to a painting of a ship.

'There she is,' he says, his fingers tapping on the glass. 'She was all lit up, bright as day. I was down here, came up No. 1 hatch. The flames were coming across from the port side to the starboard side. That's the lifeboats we tried to get away but we couldn't, then I thought if I could grab on to the railing and hold on, but I got sucked straight back down ...'

He taps the middle of the painting. 'The torpedo hit there. That's where the nurses were. I could hear them screaming. From the burns ...'

For a few seconds, Pash is silent. He turns away from the wall and looks around the room, eyes wide, voice choking and lips trembling.

'I still hear them, you know. I suffer. At night I hear them ... the nurses ... screaming ...'

***

Postscript

The wreck of the AHS *Centaur* was found by David Mearns and his search team at a depth of more than 2000 metres about 30 nautical miles from the southern tip of Moreton Island off the Queensland coast on December 20, 2009. The ship's location was less than one nautical mile from the site calculated by Gordon Rippon.

'It is timely to take a moment to remember those lives lost on that tragic day in 1943, and the devastating effect this has had on their families,' said Premier Anna Bligh on the day news of the discovery was announced. 'This morning's confirmation that the wreck has been found will hopefully provide some peace to the loved ones of those killed.'

In 2014, *Centaur* Association president Richard Jones told the BBC's Philippa Fogarty that Mearns' discovery had provided relatives with a moment of 'gratitude, one of relief and one of joy and sadness'. Jones' uncle, Gordon Jones, was one of those who perished. 'Everyone found some comfort in knowing where their relatives were,' he said.

A commemoration service was held at sea directly above the remains of the *Centaur*. 'There is always a vacant place, but all these things are huge steps forward in the healing process,' said Jan Thomas, who attended the service.

A memorial plaque was lowered to the wreckage. The *Centaur* is now protected by the Australian government under the Historic Shipwrecks Act of 1976.

Chapter 9

WAR GAMES

IT IS THE NIGHT of June 11, 2008, and as it is whenever a State of Origin match is telecast, the TV at the neat little house in Mill Street, Charters Towers, is on full volume.

Ernie Hill, 87, suffered hearing loss serving as a machine-gunner in the Pacific during World War II, but nothing can stop him from savouring his great love for rugby league. Club games, internationals and especially Origin, if it's on the telly Ernie watches it, with the sound reverberating off the walls.

On this night the commentators seated high in the stand at Brisbane's Suncorp Stadium are talking about the first Origin ever played, and how far the concept of players representing their home state, rather than the state in which they play club football, has come since. 'The first game was 28 years ago, on this very ground,' one says.

Ernie sits up in his chair and raises his voice over the din.

'No, it wasn't!' he shouts at the screen. 'It was on Bougainville in 1945. I should know, I was there … and I've still got the program.'

On June 5, tens of thousands of Queenslanders will make their annual pilgrimage to Suncorp Stadium in Milton, just west of the Brisbane CBD, for the first match in the State of Origin series.

Many will follow a well-worn path. They will walk northwest down Caxton St from Petrie Terrace, take a shortcut across the grass verge along the front of the stadium and line up to have their photographs taken alongside the statues of four of Queensland's Origin greats: Darren Lockyer, who played a record 36 games in the maroon jumper; Wally Lewis, regarded as Queensland's undisputed king of Origin; Arthur Beetson, the man credited with kick-starting it all; and Mal Meninga, who left an indelible Maroon mark on and off the field.

From the requisite photo-op, the Maroons faithful, many with painted faces and proudly wearing jerseys emblazoned with the names of their heroes on the back, will move further up the pathway for a slow stroll past the commemorative plaques embedded in the concrete concourse.

Every player who has taken the field for Queensland in an Origin match since Beetson led a team of local youngsters and Sydney-based expats to a win over the hated Blues of NSW on July 8, 1980, has a plaque on the walk of fame.

For younger fans it is the names of players like Billy Slater, Johnathan Thurston and Greg Inglis that attract the most interest.

Those with a sense of history prefer to pay homage to 'The Originals', like Beetson, Lewis, Meninga, Rod Reddy, Johnny Lang, Chris Close, Kerry Boustead and Rod Morris.

They are all there, the superstars, the characters and the quiet achievers whose names will be forever linked to the multimillion-dollar TV ratings phenomenon that is Origin: Alfie Langer, Gene Miles, Fatty Vautin, Trevor Gillmeister, Gorden Tallis, Petero Civoniceva.

Those who have played 20 games and those who have played just one; every Queenslander who has earned lifetime membership in the state's most exclusive club, the FOGs — Former Origin Greats.

Yet there are 13 names that won't be read. Names such as Jack Barnes, Bobby Williamson, Jim Christopher and Hec Bradshaw. They don't have a plaque on Origin Walk and only one warrants a mention on the nearby wall that bears a list of every player who represented Queensland before 1980.

But on September 16, 1945, at a ground in Torokina, Bougainville, they did represent Queensland against NSW in what rugby league historians consider the game's first-ever State of Origin series.

IN 2011, QUEENSLAND RUGBY league stalwart Kevin Brasch received a phone call from Major John

Wright, curator of the military museum at Brisbane's inner-city Victoria Barracks. Retiring as chairman of the Brisbane Rugby League in 2007, Brasch remained active in the game as a member of the QRL History Committee.

Major Wright told Brasch he was updating the museum's inventory and was seeking information about a trophy, made from a Japanese artillery shell, that had been awarded to the winner of a two-match 'Origin' series played between Australian troops on the island of Bougainville, about 1500km northeast of Cairns, days after the end of World War II.

The trophy was engraved with the names of the players in both teams, and Wright was seeking to trace any relatives.

The call proved timely. The QRL History Committee holds regular public lectures covering colourful and little-known chapters of rugby league's past. Committee secretary Greg Shannon, 38, was scheduled to give the next lecture, in May 2012. His subject: Rugby League in World War II.

'Kev told me about the trophy, and we went up to the barracks to see it and copied down all the names,' he says.

'My talk was about rugby league in all the armed forces. At the end I told a story as an example of the game in each of the services: navy, air force and army. I used the Origin series in Bougainville as the army example. It wasn't the major part of the lecture, but it has turned out to be the most interesting.'

Shannon, an agronomist working in the sugar industry at Tully, North Queensland, found that rugby league was the odd game out when Australia went to war in 1939.

Rugby union, Australian rules and cricket were the official sports of the Australian armed services. It was only through the lobbying of James Larkham, state Member for Rockhampton and a former president of the QRL, and Ingham-born Arthur Fadden, briefly prime minister in 1941 and inaugural president of the North Queensland Rugby League, that rugby league was adopted by the military.

Matches between battalions became regular events. There were even 'interstate' clashes between Queensland and NSW battalions, although just like the criterion that led to the introduction of State of Origin in 1980, selection was based on place of enlistment, rather than place of birth.

The most famous example is South Sydney great Jack Rayner, who was born at Coraki in Northern NSW but joined up in Queensland and served with the 61st Regiment — the Queensland Cameroons — in New Guinea. He was playing a match for his regiment in Port Moresby against the 30th Regiment from NSW when he was spotted by former South Sydney player and coach Eric Lewis, who suggested he try out for the Rabbitohs after the war.

'If I get out, I'll come and see you,' Rayner reportedly said.

Rayner captain-coached Souths to five premierships, represented NSW 16 times and played five Tests for Australia between 1946 and 1957.

RAYNER, WHO WAS STILL in New Guinea when the war ended, would never get the chance to play for his home state during his military service. The honour of playing and watching the first 'Origin' match went to those who were based in Bougainville on August 21, 1945.

With the war ending 12 days later, the Australian Army was faced with the massive logistical task of bringing its forces — including an estimated 30,000 on Bougainville alone — back home.

'They were pretty much stuck there for months with nothing to do,' Shannon says.

'The American ships took the Yanks home first and the Aussies had to wait for them to come back. The officers were doing anything they could to keep them occupied. There were cricket games, stage shows, Aussie rules and rugby league.'

In true military fashion, everything had to be done by the book, and a Bougainville Rugby League Association was formed, with Brigadier General LG Binns, later manager of the Brisbane City Council Transport Department, as patron. Lieutenant Tom Pedrazzini, a long-time official of the Brisbane Rugby League before the war, was president, and Warrant Officer Ron Connor secretary.

Matches were played on the site of a former US medical company hospital and evacuation centre, renamed Medco Ground.

Throughout the war Aussies from all over the country fought alongside each other without any thought of where they were born or grew up. They were all in it together. But with the fighting over and the men's time taken up with playing and watching sport, interstate rivalries resurfaced.

Just as their sons and grandsons would 35 years later, the Diggers on Bougainville became frustrated at seeing teams representing NSW battalions fielding players who grew up in Queensland, and vice versa.

The seed of the first Origin series was planted.

OVER THE YEARS SINCE 1980 there has been much debate and conjecture over who deserves the title 'Father of Origin'.

South of the border, Kevin Humphreys, the NSW Rugby League chief executive of the time, is given credit for starting Origin. In Queensland, then QRL chairman Senator Ron McAuliffe is recognised as the man who brought the first series to fruition, and in 1998 *Courier-Mail* journalist Barry Dick revealed it was Brisbane Broncos' founding director Barry Maranta who adapted a concept used by the Victorian Football League in 1977 and presented it to McAuliffe.

Nowhere is the name Arthur Titley mentioned.

In September 1945, Major HA 'Tiger' Titley, 36, a member of a prominent Charters Towers retail family, was a senior officer on Bougainville. When Warrant Officer Ron Connor came to him requesting permission to organise an Origin series, he was talking to the right person.

'My father was a real rugby league man,' says Titley's son Rob, who lives in the North Queensland town, 134km inland from Townsville.

'He played hooker for Rovers club in Charters Towers before the war. That's when he got his nickname, Tiger. He played his last game in 1962 when he was 53. It was the 50th anniversary of Charters Towers High School and the "Old Boys" played the "Really Old Boys". The busiest person on the field that day was the first-aid man. The only thing he had in his kit was a bottle of rum.'

Major Titley gave the go-ahead for the game and arranged for the 31/51 Battalion band to provide entertainment.

Commander of the 3rd Infantry Division, Major General William Bridgeford, commissioned a trophy that was made in the army workshop and selectors from each state started choosing their teams.

At 1400 hours on September 16, 1945, the Queensland side, wearing maroon jerseys and captained by Rockhampton fullback Jack Barnes, ran onto Medco Ground and lined up against NSW, wearing blue and led by Newcastle five-eighth Horrie Marjoribanks.

There is a certain irony about Marjoribanks captaining NSW. He was the uncle of renowned 15-Test five-eighth Bobby Banks who, though he was born and raised in Newcastle, NSW, played 26 times for Queensland and was part of the 1959 side that was the last Maroons team to beat the Blues in a series prior to Origin.

'I played against Uncle Horrie once,' recalls Banks.

'I was 17. I don't know how old he was, but my dad was the oldest of ten boys. I was playing for Central; he was playing for Wests. He was a wonderful player, a great tackler. I remember when I was in the under-16s watching him play for Newcastle against Wollongong. He was up against a bloke called Johnny Hawke and he played all over him. Hawke got picked in the Country Firsts ahead of Horrie and went on to play for Australia. I might be biased, but I reckon Horrie was hard done by. He could have been the first of our mob to play for Australia.'

The best-known of the Queenslanders was 29-year-old Sergeant Kelly Brennan, who started his career with Ipswich Brothers in 1933 and went on to play with Rialto and West End. A tough, solid hooker, Brennan played 12 matches for Queensland between 1946 and his retirement in 1948.

Other than the names on the trophy and a belief that two matches were played and won narrowly by Queensland, Greg Shannon was initially unable to unearth a great deal of detail about the series, but his lecture stirred interest, leading to a number of discoveries.

While combing through old programs, Shannon found a small story about the handover of the Japanese artillery shell cup to the Queensland Rugby League by the army at half-time in a 1946 Brisbane club match.

The article confirmed that the matches were both won by Queensland — 10–9 and 20–13.

The most valuable breakthrough was by NSW-based league historian David Middleton, who came across an article from Rockhampton's *Morning Bulletin* of September 27, 1945.

Headlined 'Football on Bougainville — Queensland's win', it was a detailed account of the first match written by Warrant Officer Connor who, as well as being secretary of the Bougainville Rugby League Association, was one of the Queensland selectors. Describing the match as 'one of the most spectacular and exciting games of rugby league I have ever witnessed', Connor said soldiers who had recently returned from leave on the mainland rated the standard of play on the island higher than A-grade.

'One man went so far as to say that this interstate match was better to watch than the one played in Brisbane a few weeks previously.'

The teams were, Queensland: J. Barnes (captain), J. Christopher, L. Ashmore, C. King, E. Lade, N. Hoare, R. Williamson, H. Bradshaw, M. Tresedor, T. Kraft, M. Thompson (vice-captain), K. Brennan, F. McLennan.

NSW: H. Parkinson, W. Peachy, D. McRitchie, T. Briggs, H. Dhu, H. Marjoribanks (captain), R. Miller, H. Taylor, V. Love, C. Smith, J. Hobson (vice-captain), D. Sinclaire, H. Freeman.

Many of the players were A-graders from Brisbane and Sydney. The referee was Brisbane's Frank Ballard and the game was broadcast to troops throughout the islands by Tom Pedrazzini.

'NSW won the toss and Bobby Williamson kicked off for the Bananalanders,' Connor wrote. 'From the word "go" the Queensland forwards were on the ball and stayed there until the final whistle.'

Winger Jim Christopher kicked two penalty goals and scored a try just before half-time to give Queensland a 7–nil lead at the break.

Midway through the second half, NSW fullback Norm Parkinson kicked two goals to move the Blues within three points and 'then kicked a beautiful field goal from about 40 yards. The kick was remarkable in that the ball hit the crossbar, bounced into the air and fell on the right side. Queensland 7, NSW 6.'

NSW then scored a try that Connor described as 'the most brilliant of the match', with centre Doug McRitchie on the end of a backline movement in which 'all men from the halfback to the winger handled'. Parkinson missed the conversion, leaving NSW ahead 9–7 with five minutes to play.

Queensland went on the attack and when Christopher missed a goal with two minutes left, it seemed the Blues were home but as is so often the case in Origin football, the final scene was still to be written.

'The Queensland forwards carried the play to ten yards from the NSW goal line and fierce rucking resulted. Thompson, who was acting as dummy half, gathered the ball and passed to Hec Bradshaw, who smashed his way through the NSW defence, was tackled by the fullback but dragged him over the line to score about 30 seconds before the fulltime bell rang. The spectators swarmed over the field and Hec was carried shoulder-high to the dressing room.'

IN JULY 2012, JOURNALIST Michael Saunders received a phone call in the office of *The Northern Miner* newspaper in Charters Towers. It was a woman who said her father had something he might be interested in looking at.

His name was Ernie Hill, she said, and during the war he'd been at this rugby league game. He still had the program … Michael drove out to Mill Street and met Ernie and his wife Mary. They told him about how Ernie had heard the commentators in 2008 talking about the first Origin and how Mary had sent off a letter to Wally Lewis telling him about Ernie's program. 'I sent it to WIN TV in Rockhampton but never heard back,' she says.

'He's a busy man, he probably never got it. Then one of our daughters rang the paper and the reporter came out here and we showed it to him. He wanted to take our picture but we were a bit too shy for that. We just thought someone might be interested, that's all.'

The newspaper ran the story and a photo of the program.

Michael Saunders had read a story by *Courier-Mail* league writer Steve Ricketts about Greg Shannon's lecture and sent Shannon a photocopy of the single-sheet program. It is now on display at the National Rugby League museum in Sydney.

Ernie Hill died in February 2013, aged 92.

'He never got to a State of Origin, but he watched every one of them on TV,' Mary says. 'He was deaf in one ear after the war and he used to have it up so loud. I think you could have heard it down the street. It didn't matter where you were in the house when he was watching football, you could hear every word. I think I knew the name of every player in Australia at one stage.'

The TV might be quieter at Mill Street these days, but the Hill family interest in State of Origin lives on through Ernie's three sons, eight grandsons and his great-grandsons. And the family still has the family heirloom, Ernie's program.

The question has to be asked: why did he keep it all those years?

'Because he loved rugby league, and he loved Queensland,' Mary says.

'It's not everyone who can say they've seen every Origin game ever played, is it? After all those years he couldn't remember who won that first game but if anyone asked him, he said, "Queensland, of course."'

Chapter 10

DAM SAN

BARRY PETERSEN LIVES IN a street that doesn't exist. The concierge at my hotel doesn't know where it is so he calls someone who lives nearby. It's like a scene from *Seinfeld*. What should take a minute or two becomes ten, 15. The concierge looks more and more confused as he scribbles down notes.

'She can't explain,' he says finally, putting down the phone and handing over a page of directions. 'Give this to the taxi driver. He'll know.' He doesn't.

Even for Bangkok, a city where markets can spill over six blocks of CBD streets at night and vanish without trace by morning, this is unusual.

The driver stops on a busy road, looks down at the instructions, up at a collection of beauty salons, offices and shops, and back down to the instructions. In Bangkok, more than in most cities, time means money and he's not wasting any more of either on this lost cause.

'There,' he says, reaching over to open the passenger door. 'It's there.'

It isn't, and no one in the shops or offices knows where it is. A phone call and a 'don't move, I'll come and get you' and he emerges, a dapper, white-haired man of 73 with a slight limp and a big smile. Barry Petersen, Dam San, chief of the Tiger Men — the Queenslander who was the main player in one of the most remarkable stories of the Vietnam War.

He leads the way down an alley, invisible from the road and ending abruptly at a high wall. On the other side stands a large building festooned with aerials and satellite dishes. Petersen's apartment is in one of the last buildings in the alley. Three levels, nondescript from the outside, it was once also his office. The framed pictures on the walls reflect his love of Asia; the neat piles of books and records the meticulous nature of a career army officer.

The apartment is comfortable but by no means lavish. One gets the feeling that while Petersen lives in Bangkok, his thoughts are elsewhere: some 1000km to the north-east, in the highlands of Vietnam. He offers a glass of water, cranks up the air-conditioning and asks, 'Where do you want to start?'

How about at the beginning, as in how does a kid from Sarina in North Queensland become a chieftain of a Vietnamese highland village, work for the US Central Intelligence Agency or command a personal army of more than 1000 mercenary soldiers?

'You have to be in the right place at the right time,' says Petersen. 'And start young.'

> When I was a boy I used to read those books by Edgar Wallace, you know, *Sanders of the River* ... books about the British Colonial Office in Africa. I actually wrote to the Colonial Office in London and said I was interested in joining. They wrote back and said they didn't employ 15-year-old boys, but they kept me on their mailing list.

While it might have been Edgar Wallace who introduced Petersen to the outside world, it was his grandfather, Hans Christian Petersen, a first-generation Australian of Danish descent, who nurtured the gumption to go out and explore it.

When Barry's dad closed down his small radio repair business and headed off to World War II, it was left to Petersen's mother to raise him and his two sisters. Through those years he spent long hours with his grandfather, a man who, he says, 'had no time for the faint-hearted'. One of Petersen's abiding memories is of wading out into deep water off Sarina Beach as a ten-year-old, his grandfather's fishing net in his hand. He was terrified of sharks but, spurred on by the old man's gruff encouragement, pushed his fear aside and completed the job. It was to be a pattern of behaviour repeated many times over the next 60 years.

Called up for national service during the Cold War of the 1950s, Petersen found army life a better fit than his traineeship as a surveyor under future Brisbane lord mayor Clem Jones. He joined the regular army and underwent officer training at Portsea, Victoria. By the middle of 1962 he had completed two years as an infantry platoon commander in Malaya, but it wasn't just his expertise leading patrols in search of CTs — Communist Terrorists — during the so-called Malayan Emergency that struck a chord with his commanding officers. It was also that, on his own initiative, he spent his leave learning the local language so he could converse with native villagers.

Soon after returning home, he became one of the first Australians to serve in Vietnam. By the time he was posted in August 1963, his literary tastes ran to Ian Fleming's James Bond and his interest in Africa moved to Asia. Within two years he would be a mix of Sanders, Bond and Colonel Kurtz, the pivotal character of Joseph Conrad's novel *Heart of Darkness* brought to life by Marlon Brando in *Apocalypse Now*. The difference between him and Kurtz being that he wasn't assassinated.

But not for want of trying.

CAPTAIN PETERSEN WAS PART of a contingent of 36 officers, warrant officers and sergeants assembled under the title Australian Army Training Team Vietnam, soon to be known simply as The Team. Between 1962 and 1972, 992 members of The Team served in Vietnam. Among them, 33 were killed in action and 122 wounded. They were awarded 147 medals for bravery and meritorious conduct — including four Victoria Crosses and a US

Distinguished Service Cross — and earned a reputation as the best of the best, but none experienced anything like what Barry Petersen did.

> I trained with that first group and expected to go with them but when the names were posted I was listed as a reserve. I was pretty disappointed. I went up to our commanding officer, Colonel Serong, and told him how I felt. He said not to worry, that sometimes things worked out. I didn't know then that he had plans for me.

As the first members of The Team headed to Vietnam, Petersen was recommended for Special Operations, a euphemism for espionage, counter terrorism and guerrilla warfare — all the things that aren't found in the regular army handbook. He took courses in jungle warfare and intelligence, also graduating from what its survivors called the School of Torture, a secret facility at Sydney's Middle Head where the chosen few learned first-hand what to expect if captured and interrogated by the enemy.

> When I got over to Saigon a year later, Colonel Serong said he'd got permission for me to work in the field. Up in the highlands with the CIA. I went to see them and they gave me the standard CIA issue — a pair of khaki drill trousers, a 9mm Browning automatic pistol to stick in the waistband, and a short-sleeved shirt worn hanging out, to hide it.

On the flight north from Saigon to the highlands in August 1963 Petersen was briefed by his CIA case officer, one of the covert operatives Australian soldiers in Vietnam would refer to as 'Sneaky Petes'. Over the previous year the CIA had been funding a program to recruit indigenous highlanders to serve in a locally based paramilitary force. There were up to one million Montagnards — the name given by the French to the many indigenous mountain tribes inhabiting parts of the Indochina peninsula — at the start of the war, and the relationship between the Montagnards and the Vietnamese people and government was uncomfortable at best.

The animosity increased after Vietnam's partition in 1954, when refugees began flooding south and the South Vietnamese started a program of resettlement in the highlands. Then in the early 1960s the Montagnards suddenly found themselves very popular indeed. Their villages could provide shelter to the Viet Cong (the guerrilla force that operated with the backing of North Vietnam) as they moved further south, or a defensive barrier for the South Vietnamese. To get them on side the Viet Cong dangled the carrot of future self-rule. The South Vietnamese, thanks to the support of the CIA, could offer something far more in the here and now: money.

In mid-1962, US advisers convinced the South Vietnamese government to recruit, train and arm Montagnards, forming them into small platoon-strength units. They would be paid depending on the work they did and the action they saw. In effect, the CIA would turn these mountain villagers into a mercenary army. In the province of Darlac, Captain Barry Petersen of Sarina became their commander-in-chief.

Darlac, some 265km north of Saigon, was home to the Rhade people, who lived in villages scattered around the capital city of Ban-Mê-Thuột, home to around 45,000 South Vietnamese. It was here that Petersen set up his headquarters and took charge of what could only be termed a haphazard enterprise. For months the CIA had been supplying arms and $2500 a month to the provincial chief to operate a force of 100 Rhade recruits, but it had yet to see any soldiers, let alone receipts. Wresting control of the operation away from the local authorities was one of Petersen's first tasks.

'The CIA liked to have foreigners do their work for them,' he says. 'That way if it all blew up they could say, "Who? We don't know anyone by that name."'

Petersen's case officer left him in Ban-Mê-Thuột with $350 cash and told him to get to work. He'd be back with more money in a few weeks' time.

'You'll know when I'm coming,' he said. 'I'll get my pilot to buzz the town.'

And so began what a senior Australian officer would later describe as Petersen's 'Boy's Own adventure'. He rented a house, hired a cook, found an interpreter and built an army.

Petersen was a man of many talents. He was an experienced and extremely competent soldier who struck an easy rapport with the Montagnards, or 'Yards' as the Australians called them. He quickly grasped their culture; in turn they took to him as one of their own, inviting him into their homes and to their traditional rice-wine drinking ceremonies. The ultimate sign of acceptance came when they named him 'Dam San' after a legendary Rhade warrior. Over two years, assisted by a succession of Australian warrant officers and his interpreter, Jut, Petersen built the ragtag collection of 100 recruits into a well-trained and highly effective force of more than 1000. He was given free access to the CIA supply depot in Saigon and had his initial $350 rolling budget increased to $50,000.

To the CIA, Petersen's units were known as Armed Propaganda and Intelligence Teams. The South Vietnamese named them the Truong Son Force, after the mountains in which they operated. Impounded Viet Cong documents revealed the enemy referred to the hit-and-run patrols that struck with deadly force and disappeared back into the jungle as quickly as they came as 'Tiger Men'.

With every day spent in the highlands, Petersen's influence grew. As well as coordinating recruitment and training, he led patrols, built hospitals and schools. The captain from Queensland found himself in an unprecedented position of power. At that time, the highest ranking Australian in Vietnam was in charge of a force of about 50; Petersen commanded more than 20 times that number.

Inevitably, he became the target of jealousy and animosity. The attempts at intimidation started within a few days of his arrival in Darlac, with two men trying to get into his hotel room, then standing and staring at him through the window. When he moved into his own house a month later the scare tactics went to a new level. At the local markets, his cook was given a freshly killed and cleaned chicken for her master's dinner. She fed it to a household pet, a leopard named Fatima, which was found dead, poisoned, the next morning.

'At first I thought it was the Viet Cong, but it could have been the local province chief,' Petersen says. 'Until I got there he was receiving cash from the CIA to fund an army that didn't exist. I was costing him money.'

For its part, the CIA saw Petersen as a useful, if expendable, operative, but when he refused to start training his Tiger Men as political assassins, as instructed by a case officer, he tested the agency's patience. The longer he stayed in Darlac, the steeper the tightrope he found himself walking. The Yards saw Dam San as a trusted insider who could help their push for autonomy through his American contacts. When Montagnard forces revolted in August 1964 and took South Vietnamese and US officers hostage, it was Petersen they called on to break the deadlock and ensure their demands were heard — a display of influence that did little to ingratiate him with the South Vietnamese or the CIA. He discovered later that a high-ranking South Vietnamese officer believed Petersen had been a CIA agent planted to instigate the revolt and had ordered his death. Only a case of mistaken identity saved him. Meanwhile, his CIA contact warned him he was becoming a 'personality cult' among the Montagnards and dubbed him, without humour, 'Lawrence of the Highlands'.

'Another Australian officer who was close to the CIA told me years later that I was lucky to get out alive,' Petersen says. 'He said my CIA contact at the time was known as someone who "took the easy way out". By that he meant he'd have people knocked off when they didn't toe the line.'

The end for Petersen came when he received a letter from a Montagnard leader exiled in Cambodia asking him to act as a go-between with the Americans as he negotiated a return for himself and his followers. What had started as a military exercise had become embroiled in politics and dirty dealings. The Montagnards were remarkably prescient when they gave Petersen his nickname: like Dam San, he upset an enemy too powerful to overcome.

In August 1965, he was ordered out of the highlands. The farewell lasted over a fortnight, culminating in a 20-hour ceremony at the main Truong Son camp where Petersen was anointed a Montagnard chieftain. He was presented with traditional clothing, now held at the Australian War Memorial in Canberra, and for the last time inspected his troops, some 500 of whom assembled to bid farewell. The Boy's Own adventure had come to an end.

OR HAD IT? PETERSEN might have upset his CIA contacts on occasions, but he was to find the shadowy world of 'Special Ops' hard to leave behind.

In March 1965, he had been recalled to Australia on unofficial leave in order to debrief the Australian Secret Intelligence Service, Department of External Affairs and the army's Directorate of Military Intelligence on the situation in Vietnam. In Melbourne, he lunched with the head of ASIS at a time when only six or seven people in the country knew his identity. Prime minister Sir Robert Menzies dropped by their table to say hello. Seven months later, after he was removed from South Vietnam, he went to Borneo for two months as a guest of Britain's Secret Intelligence Service, MI6.The British were carrying out similar operations in Sarawak and Sabah to those of the CIA in Vietnam, using ethnic minority tribespeople in covert campaigns against Indonesia. Petersen was given carte blanche to inspect the top-secret procedure.

Over the next 20 years he would study psychological warfare at the US Army Special Warfare School at Fort Bragg, return to Vietnam in 1970 for 18 months commanding C Company 2RAR out of Nui Dat, be seconded to the Malaysian Army for two years, and serve in a joint ASIS-MI6 operation working with guerrillas in New Guinea. After his retirement as a lieutenant colonel in 1979, he was called in by the army to lecture 'younger guys' on special operations at a training centre outside Cairns. He was also involved in at least two undercover stings on behalf of Australian authorities. And then on to the latest chapter, set in Bangkok.

'I'd always had a dream to retire to a rural property in the tropics and when I left the army I bought a place in Cairns,' Petersen says. 'It didn't take me long to realise that I wasn't made to sit around. My life lacked mental stimulation, so I started a landscaping business and that's what brought me back to Asia. I was importing garden ornaments, which meant business trips back. After a while I realised my heart was still in Asia. In 1990, I was visiting Cambodia and one night I looked out over the Mekong River and said, "That's it. I'm coming back."'

As he had never married or had children, it was not difficult for Petersen to begin a new life in Bangkok. For a few years he ran a business helping international companies establish themselves in Thailand, guiding them through the bureaucratic maze. Although he eventually withdrew from the business, he remains close to the Thais he employed. He sees them most days, sharing meals and shopping trips.

'I think of them as my family,' he says. 'This is a wonderful country to grow old in. Young people here don't just ignore you or push you aside.'

There is more to it than that, of course. The office in Petersen's apartment is given over to files, maps, letters and newspaper clippings concerning the ongoing persecution of the Montagnards since the fall of the South Vietnamese government in 1975. The highland people still long for freedom, and they are still treated as 'moi', or savages, by the ruling party. According to US-based Human Rights Watch, more than 350 Montagnards are being held in Vietnamese jails on charges such as 'undermining national unity' and 'causing political instability'. One, Y-suan Mlo of Ban-Mê-Thuột, was sentenced to ten years' jail for using a cell phone to contact relatives in the US.

On that trip to Asia in 1990, Petersen returned to Ban-Mê-Thuột for the first time since the war under the guise of being a former Australian war correspondent. His cover was blown when old villagers, recognising their beloved Dam San, started grabbing his hands and kissing them. The trip was abruptly ended, and the villagers were arrested and interrogated by the Vietnamese secret police.

Petersen continues to lobby for the Montagnards and takes an active interest in their welfare. He had made an unsuccessful attempt in 1975 to sponsor his former translator Jut and his family as immigrants to Australia. Flown out of Vietnam by the CIA as the country fell to the communists, Jut made it to the US, where he still lives. The two have stayed in irregular contact, but Petersen's involvement with Montagnard refugees doesn't end there.

While he is giving nothing away, it would appear obvious that Bangkok is a better place than Cairns for an old 'Sneaky Pete' to keep in touch with goings-on in Vietnam and take an active hand. Petersen won't elaborate, but there are veiled references to one case involving an assumed identity, false papers and a late-afternoon crossing at a crowded but relatively remote border checkpoint to try to assist one escapee.

'I feel we have a responsibility to help them,' he says. 'I feel very upset about the way we have abandoned these people. We've sold them down the drain. I just do whatever I can.'

Just how much he has done, or will do, he won't say. Friendly, articulate and seemingly open, Petersen is excellent company. He loves to talk about his past and will tell you everything he wants you to know — and not one word more.

Now he leads the way downstairs and into the heat of the Bangkok afternoon. The large building covered in aerials and satellite dishes seems to stand sentinel.

'You know what that is?' he asks. 'That's CIA headquarters."'

It seems too pat to be coincidence. The question has to asked. 'You're not still working for them, are you?'

'Who, me?' he half-laughs. 'I'm just a harmless old retiree.'

And in the blink of an eye he's gone. Vanished into a street that doesn't exist.

Chapter 11

THE LOST BOY

WHEN FIRST HEARING THE story of an 11-year-old Australian boy executed by the Japanese as a spy in World War II, one can't help but think two things: this is shocking, and how have I not heard of it before?

When he stumbled onto details of the firing squad death of Dickie Manson and his mother Marjorie in May 1942, Brisbane author Ian Townsend, 56, thought exactly that. The result is his book *Line of Fire*, published by Fourth Estate.

Author of two works of fiction, *Affection* (2007) and *The Devil's Eye* (2008), both of which told human stories set in times of national disaster, Townsend initially planned to follow a similar path with his third book — invent a family, put them in a wartime setting and build a plot around them. And then he discovered Dickie Manson.

Born in Melbourne, educated for a short time in Brisbane at Fortitude Valley Boys' School and then taken by his mother to the then Australian territory of New Guinea a year before the Japanese invasion, Dickie was court-martialled and executed soon after his 11th birthday. For 75 years, the facts behind his death were a painful mystery to his relatives back in Australia and a long-forgotten footnote to one of the most shameful episodes in Australian military history. If not for Townsend's curiosity and research skills, they would have remained that way.

As a former ABC journalist with a particular interest in science and history, Townsend saw Rabaul, at the north-eastern end of the island of New Britain, as the perfect setting to combine the two. Famous for its volcano and spectacular harbour formed in a volcanic caldera — or crater — after an eruption about 1500 years ago, Rabaul was also of major strategic value during the war, leading to its bombing and subsequent occupation by the Japanese in January 1942.

'I'd always wanted to write a book about 1942 and set it in Queensland,' says Townsend as he leans back in a lounge chair at his home in Brisbane's northwestern suburb of The Gap.

> When you listen to people talk about the war years on the home front in Australia, it seems like there was something almost attractive about that time; it was like they felt quite alive with the ever-present danger of the Japanese, who everyone thought would try to invade. I like the idea that when people feel

> they are at risk, a lot of interesting things come out, so that's why I thought of 1942. Then I started looking at Rabaul as a result of that, because during the bombing up in North Queensland a lot of the planes were coming from there.
>
> The more research I did about the Battle of Rabaul, the more I wondered why people don't know much about it. So basically I thought I'd do some research and set my book in Rabaul … and then I came across this family.

Once he unearthed snippets of information about the trial and execution of Dickie, Marjorie, 29, her partner Ted Harvey, 54, Marjorie's brother Jimmy, 27, and plantation manager Bill Parker, 43, any thoughts of another novel were scrapped. The truth was far stranger, and much more compelling, than fiction.

> The fact that they were shot was in some of the military histories of Rabaul, but there wasn't much about it, just a few paragraphs, and this family was so intriguing I couldn't fictionalise it. I felt I had to find them.

Rather naively, he now admits, Townsend thought this would be a relatively easy task: just track down relatives in Australia and ask them everything they knew. It proved anything but simple. Manson was Marjorie's maiden name. Unwed at the time of Dickie's birth, she married his father, Jack Gasmier. Later, she left Gasmier and moved to Rabaul to be with copra plantation owner Harvey, taking his name during her time in New Guinea.

> I ran into a brick wall because they changed their names so many times through the course of their lives — the little boy had three surnames in his short life — so tracking down all the references to these different people was very difficult. It became a challenge in the end, but as a journo, if there is a barrier there you want to go past it. It became a personal thing. I was obsessed. I thought, why don't I know the story about Rabaul? Why don't I know about this family who were executed for espionage and buried beneath a volcano? There was a three-day trial: why don't I know this? I thought, this is interesting to me, but will it interest other people? Then I thought, bugger it, I'm interested, I'd like to read a book about this, so I'll just pursue it.

The chilling facts behind what Townsend believes is the only military court martial and execution of an Australian child were not too hard to piece together. The Australian government, knowing the Japanese were on their way to Rabaul, sent a ship to evacuate all women and children, while leaving the civilian men and the military garrison to face the enemy. Marjorie, fatefully, decided to stay. When the Japanese invaded, Marjorie, Dickie, Jimmy,

Ted and Bill fled into the jungle. Betrayed by a local man, they were caught and searched. Ted had a two-way radio with him, with which he had been trying to contact Australian aircraft. Marjorie had hidden a pistol in her bag.

They were court-martialled, found guilty of espionage and, on May 18, 1942, driven in the back of an open truck to the base of Tavurvur volcano. Marjorie tied a strip of red material, cut from a native lap-lap, around Dickie's head to cover his eyes. They were led from the truck and lined up. Marjorie held one of Dickie's hands, Ted Harvey the other, and on the order of Japanese naval officer Mizusaki Shojiro, a firing squad of six soldiers shot them dead.

That much could be gleaned from eyewitness reports given to Australian authorities investigating possible war crimes after the war. For the real story of Dickie Manson and his mother, who they were and how they came to be buried under a volcano in New Guinea — and through them the story of Rabaul and Australia in 1942 — Townsend had to dig deeper.

His obsession took him to national archives around the country, old military files and newspapers and on a trip to Rabaul. The breakthrough came when he found a reference to Marjorie's brother George, who had been in Rabaul but joined the volunteer militia and fought alongside Allied troops when the Japanese landed. He sought refuge in the jungle for about three months before being evacuated by the military to Australia.

> I didn't know George existed when I started looking because he wasn't mentioned in the military histories, but when I came across him I thought, did he survive the war? I started going through old newspapers online, and his name popped up. He was engaged after the war in Adelaide, so then it was 'great, he survived the war, maybe he had kids'.

Townsend found a newspaper article about a reunion of people who had been in New Guinea before the war. George had attended, but the eureka moment came when the article included a quote from a Beverley McLean in which she said, 'My father, George Manson …'

And so the search continued. McLean is a common name, and Beverley and her family lived a long way from Adelaide, in northern NSW, but finally Townsend found her and, through her, Marjorie's third brother Graham, 94, the only person still alive who could tell him first-hand about Dickie and Marjorie. Townsend spoke to Graham Manson at his home in Sydney.

> He is still grief-stricken today. He finds it very difficult. He can talk about them before the war but if you start to approach the war with him he becomes very upset and he physically doesn't seem to be able to speak. He wants to but he can't.
>
> He didn't know much about what happened in Rabaul himself because all that was kept from the family by authorities, so I wanted him to describe

what Marjorie and Dickie were like before the war. I needed that personal information so I could paint a picture of them because so little remains. They live in him.

Townsend believes the Australian government suppressed details of what happened to the civilian men left behind in Rabaul — most of whom died — because of shame over not evacuating them when it had the chance. The lack of information shared within the Manson family was due to the most powerful of all emotions: grief. Marjorie's mother Phyllis was shattered by the deaths of her daughter, son and grandson. In her pain she destroyed any documents, personal letters or photographs that reminded her of her loss. In 1956, she committed suicide at the age of 65.

When I gave the book to Bev she read it a couple of times and she rang me up and said, 'For the first time I know who they are,' and that was great. It's nice to help people. The war crimes case was dropped so the family never really got justice, but I think this story being told means anyone who is interested can have an understanding of what happened, and that's important.

Even though they were just a tiny part of the war, through them we can see a bigger picture. They were just an ordinary family but people matter. We need to know that they existed.

Chapter 12

IN THE NAME OF THEIR FATHERS

AS 21-YEAR-OLD JACK ANNING looked across the Libyan desert in the early morning light and saw the Panzer tanks coming toward him, with elite German soldiers either riding on the tanks or fanned out jogging behind them, he felt no fear.

'I was wildly excited,' he says. 'I was sick of running. We all were. We were anxious to have a go.'

It was Easter Monday, 1941, on the outskirts of Tobruk, Libya's Mediterranean port town near the Egyptian border. Over the next few hours Anning and the rest of A Company from the all-Queensland 2/15th Battalion would indeed 'have a go' at Lieutenant General Erwin Rommel's Afrika Korps. In their first experience of combat, they would take on one of Rommel's favourite commanders and his battle-hardened troops.

By mid-morning, the fighting was all but over. The German commander lay dead in a ditch, his body covered by a bloodied Nazi flag, as his distraught men, many in tears, surrendered their weapons to the Queenslanders.

It was a crucial element in a victory that ensured the continuation of the eight-month-long siege, leading to the enduring legend of the 'Rats of Tobruk' and the eventual derailment of Rommel's march to the Suez Canal. Yet until 2011, the only people who knew about it were the ones who were there and the family members they told. The action of the men of the 2/15th during Tobruk's Easter Battle was lost under the desert sands. It was left to their sons to rewrite history.

John Mackenzie-Smith remembers well March 2, 1943, the day his father Captain Greig Smith came home from the war. John Smith (the 'MacKenzie' was added in 1989) was six years old. His father, the officer commanding A Company, 2/15th Battalion, was a day off his 31st birthday. John was playing in the front yard of the family home on Waterworks Road, Ashgrove, six kilometres west of the Brisbane CBD. A car pulled up and a man dressed in a light-coloured khaki uniform got out.

'Are you my daddy?' John asked.

'Yes. Where's Mummy?'

'At the butcher.'

'You'd better go get her.'

The boy raced to tell his mother the news. She took his hand and ran back down the street, calling through the door as she passed each shop, 'My husband's just back from the Middle East. My husband's just back from the Middle East.'

With serious injuries sustained in Libya ruling out a return to the front, Captain Smith brought back souvenirs from Tobruk that would fascinate John and his younger brother Raymond for the next 70 years: a German combat helmet, a Luger pistol, a pair of field glasses, a map titled 'Tobruch Defences' and a bloodstained Nazi flag.

Other sons had their own memories. William Yates' father, Lieutenant Ron Yates MC, was the officer who accepted the German surrender on that Easter Monday. Mal Scarr's dad Bob was on one of two Bren carriers — light armoured vehicles fitted with a Bren machine gun — that strafed the enemy position. Before his death in 1969, Sergeant John 'Bomber' Neal (later the Regimental Sergeant Major) would tell his son John of how he shot his rifle into the gun barrels of the advancing tanks in the hope of jamming them. Greg Snow's father Russell was a member of 2/15th Battalion headquarters company.

Unlike the others, Steve Rowan never saw his father return from the war. A member of A Company at the time of the Easter Battle, Sergeant Tony Rowan was killed three months later when he stepped on a 'Jumping Jack' landmine outside Tobruk. It was the day after Steve's fourth birthday, but his emotional ties to his father are as close today as they were then. On his right arm is tattooed the 2/15th Battalion's colour patch, a purple T for Tobruk, and underneath, the words 'Son of a Rat'.

Growing up, they all heard the story of the battle. They heard it from their fathers or their fathers' comrades-in-arms. They read their diaries and listened as the old soldiers talked on Anzac Day or at reunions organised by the 2/15th Battalion Remembrance Club. They heard how after days of fierce fighting, the Germans broke through the line held by the NSW-formed 2/13th and 2/17th battalions. How the 120 men of A Company dug in 100 metres in front of the guns of Britain's 1st Regiment, Royal Horse Artillery, to form a last line of defence. How, despite heavy casualties, the British gunners sent the Panzers packing while the Queenslanders held firm.

They heard how Greig Smith then ordered Ron Yates to lead his 30-man platoon on a counterattack against the remaining German infantry of the 8th Machine-Gun Battalion, which had supported the Panzer assault; how, although well outnumbered, they killed the German commander and captured 100 of his men. And they heard how no one ever gave them any credit for it.

The 'oversight' that saw the 2/15th written out of history started with the published account of the siege of Tobruk by respected ABC war correspondent Chester Wilmot in 1944. Wilmot, who died in a plane crash in 1954, wrote that the 2/15th had been stationed in reserve at the 'Red Line', Tobruk's outer line of defence, held by the 2/13th and 2/17th battalions, some 3000 metres in front of the British artillery, which were on the 'Blue Line'. He claimed that after breaking through the Red

Line the German tanks advanced 2000 metres before being turned back by the guns, and gave credit for the final operation in which the remnants of the supporting German infantry were defeated and captured to the two NSW units.

At the time Wilmot was putting the finishing touches to his book in Sydney, members of the 2/15th had more pressing matters than recognition on their minds — they were fighting the Japanese at Kumawa, New Guinea. It was only later, after the war, when one publication after another rehashed Wilmot's version of events that they began to feel aggrieved. When they got together, they talked about setting the record straight but no one knew how. They left it until the next reunion and the next, and then time began to run out.

When the battalion's Remembrance Club was formed in 1946 it had several hundred members, nearly all returned servicemen. Sixty-five years later, most members were widows, children, grandchildren and great-grandchildren of the original 2/15th. Just 48 World War II veterans remained; 16 of them Rats of Tobruk, including club president Gordon Wallace, and only three — Doug Smales, Eddie Stott and Jack Anning — survivors of A Company. The story of their part in the Easter Battle was at risk of dying with them.

'The passion was there,' says Steve Rowan, secretary of the club since 2002. 'We wanted to get the recognition for our fathers and grandfathers, but none of us had the expertise. We didn't know where to start.'

Fittingly, it was John MacKenzie-Smith, the son of A Company's commanding officer at the time of the Easter Battle, who led the way. A retired senior lecturer in teacher education, Dr MacKenzie-Smith, 75, is now a respected historian, but until 25 years ago he had little interest in military history. It was his father's death from cancer in September 1987 that sparked MacKenzie-Smith's curiosity over the bloodstained flag he had brought back from Tobruk, eventually leading him to uncover the truth of the Easter Battle.

'My father would tell my brother and me what had happened over there if we pressed him,' he says. 'I basically knew by the age of seven what the story was. It was to do with the concrete things he brought home: the German flag, the German helmet that we used to goosestep around the yard with, the Luger pistol, a pair of binoculars and an Italian map of Tobruk defences. These were his pride and joy, so early on we learned he was in a battle at Tobruk which also involved tanks. He told us they took on the German infantry that arrived there and defeated them and took 100 prisoners, but the leader of the infantry was killed. Because he was their commander, the Germans draped the flag over him. Then, after the battle was over, Aussies being Aussies, Dad's men whipped the flag off the German and gave it to my father. It was like they had captured the enemy's standard and presented it to him as their commander.'

The word-of-mouth stories were MacKenzie-Smith's only link to the action of that Easter Monday until 1976, when his father gave him a copy of the June edition of the Remembrance Club magazine *Caveant Hostes* (the battalion motto, from the Latin for 'Let Enemies Beware'). It included a detailed description of A Company's involvement on the day by Company Sergeant Major Kevin Robinson, who led one of the two sections involved in the counterattack. It backed up Captain Smith's version of events and added another layer to the story of the flag.

Detailing how the Germans had taken cover in a 'half-dug tank trap', he wrote: 'I believe one of the Bren gunners in Lieutenant Yates' group shot the German Commander, a Lieutenant Colonel, through the neck and killed him, and that seemed to demoralise the Jerries. They covered him in a big swastika flag.'

As his father lay dying 11 years later, MacKenzie-Smith brought him a copy of the recently released official Australian War Memorial history of Tobruk and El Alamein written by former barrister and World War II officer Barton Maughan. The section on the Easter Battle gave scant mention to the 2/15th.

'He didn't say why the action was important or the circumstances,' MacKenzie-Smith says. 'Most of all, he didn't say where it was and that was vital, because that was the thing that wrote them out of history for nearly 70 years.'

The book did achieve one thing, though. It started MacKenzie-Smith and his father talking about the flag. 'I said to him, "Dad, who was under that flag?" and he said, "I don't know. He was just another bloody Hun to me." That might shock people today, but that encapsulated the way they took on Rommel. They took him on face value, not on reputation.'

MacKenzie-Smith says for nearly three years after his father's death he put thoughts of the flag 'to the back of my mind' as he taught and studied for degrees, but eventually his curiosity got the better of him. 'It became my quest. I wanted to find out who was under that flag.'

With Sergeant Major Robinson stating in his *Caveant Hostes* article that the felled German was a Lieutenant Colonel, and knowing the action occurred on April 14, 1941, MacKenzie-Smith combed published reports on the battle in search of a match. He found it in Maughan's official history. The officer could only be Lieutenant Colonel Gustav Ponath, a recipient of the Knight's Cross, Germany's highest battlefield decoration, and commander of the 8th Machine-Gun Battalion.

The site and circumstances of Ponath's death have long been cause for conjecture in both Germany and Australia, and his remains have never been found. The official war diary of the 8th Machine-Gun Battalion states that Ponath was killed on the Red Line, 3000 metres south-west of the 2/15th position. 'That war diary cooked the books,' MacKenzie-Smith claims. 'It wasn't going to say their men were captured ignominiously or that Ponath died in a ditch. It had them on the frontline, valiantly trying to make their way back to their base when Ponath was shot by a member of 2/17th battalion.'

MacKenzie-Smith wrote to the Australian War Memorial about his theory, 'but they thought it was just Joe Blow and ignored it'. Researching in July 2009, he came across 'Axis History Forum', an online discussion board for those interested in German military history. In response to a member's request for any information about Ponath, MacKenzie-Smith gave a brief description of his death, including Robinson's view that Ponath was 'cut down by a Bren gun burst to the neck'.

By chance, Toowoomba ophthalmologist Dr William Yates, son of Lieutenant Ron Yates who had died in 1975, happened to read the posting. 'I was surfing the net and I came across this entry about Ponath,' Yates says. 'I was interested because I recall as a boy Dad telling me that he had killed Ponath. He said that the Germans had been trying to surrender and Ponath had run back into the trenches and started firing, and my father had spun around with his pistol and shot him.

'Dad felt sorry about Ponath. He had a lot of respect for him. We always knew it was Ponath because Dad used his name. He'd gone through his wallet and found a photo of his wife and son, who would have been about the same age as my older brother. Dad always said he intended to write to Ponath's widow but never did.'

Yates replied to MacKenzie-Smith's posting, giving his version of events and the two men began corresponding. It was Yates who urged MacKenzie-Smith to focus on telling the story of the 2/15th involvement in the Easter Battle for once and for all.

'It was such a big event in Dad's life, but it was like it just disappeared off the pages of history,' he says. 'John being a historian, I said, "It's up to you to correct this or the true story will never be told." Luckily, he decided to take up the challenge. It's just sad that our fathers weren't alive to see it.'

In June 2010, MacKenzie-Smith began researching in earnest. The history of the 2/15th Battalion up until Easter Monday 1941 was relatively easy to access. The first Queensland battalion of World War II, it was formed in January 1940 and just over a year later was part of the Australian 9th Division sent to North Africa, where it came under the command of Lieutenant General Leslie Morshead. Its mission was to relieve elements of the 6th Division, which had been dispatched to Greece following the Allies' crushing defeat of Italian troops in Libya. Crucial to that victory was the capture of the heavily fortified harbour port of Tobruk, a vital source of supply by sea.

The Queenslanders arrived in Libya in early February and headed to the frontline near Mersa Brega, 490km southwest of Tobruk. At the same time, Rommel and his Afrika Korps were moving north in a bid to regain ground lost by the Italians and then push on to Egypt and the Suez. Not equipped for a confrontation with Rommel, Allied Command ordered a tactical withdrawal from Libya. The Allies' plan was to make a stand at the border with Egypt, but they needed time to prepare. General Morshead's 9th Division was ordered to dig in at Tobruk and hold off Rommel for as long as possible. Allied Command hoped for 30 days. Morshead and his 'Rats' gave them 240.

Battered by blinding sandstorms, under attack from German aircraft and with Rommel on their tail, the Allied forces endured a nightmarish two and a half weeks on the run from Mersa Brega before making it to Tobruk and digging in on April 9, 1941. The 2/15th was particularly hard hit. On April 7, after a two-hour battle, its entire headquarters company, including commanding officer Lieutenant Colonel Robert Marlan and his second-in-command Major Charles Barton, plus five more officers and 150 rank-and-file, were captured by the Germans. After the war, those who'd been taken prisoner that day spoke of the well-drilled outfit that surprised them in the desert, and its dashing Commanding Officer. It was Lieutenant Colonel Gustav Ponath and the 8th Machine-Gun Battalion.

The Siege of Tobruk began on Good Friday, April 11, 1941, when Rommel launched his first attempt to break through the Red Line outer defences. Three days later, after fierce fighting during which the 2/17th's Jack Edmondson earned the first Australian Victoria Cross of World War II, Ponath and his troops broke through the line and headed towards the guns of the Royal Horse Artillery.

And that's where things got murky. Every document MacKenzie-Smith found had the 2/15th back in reserve, out of harm's way as the Brits and New South Welshmen fought. Even the official 2/15th war diaries were inconclusive. Yet he had heard his father's stories, read Robinson's account and now had William Yates offering his own father's recollections. Most of all, he had the flag. How could his father's men have possibly taken it from Ponath's body if they hadn't been involved in the action?

It was then he came across his father's papers and what he calls 'the bingo element'.

'That was the breakthrough that was going to put them back into history,' he says.

Sitting in the comfortable lounge room of his home at Hendra in Brisbane's inner north, MacKenzie-Smith hands over an official-looking map. Captioned 'The Easter Battle 1941', it clearly shows the Red Line marked with crosses designating Australian infantry and an arrow pinpointing the German attack. Some 3000 metres to the north, above the words 'Blue Line', are four crosses sitting inside semi-circles, designating artillery. Alongside them, someone — presumably Captain Greig Smith — has written 'R.H.A.' for Royal Horse Artillery and just below, between the guns and the enemy, 'A Coy 2/15'.

As well as the map, Smith's papers held another clue — a photocopy of a special 'order of the day' sent out by Major General John Lavarack, Commander of the Western Desert Force, on the evening of Easter Monday, 1941. In it Lavarack praises 'all ranks ... on the stern and determined resistance offered to the enemy's attacks with tanks, infantry and aircraft today'. He makes particular mention of 'a prompt counterattack by reserves of the

20th Bde'. The words are underlined, and alongside, Smith has written 'Our Coy'.

MacKenzie-Smith pushed ahead, alerting the Remembrance Club of his work. The sons of the Rats answered the call. William Yates provided his father's diary. Mal Scarr arranged an interview with his father Bob who, a week before his death, shared his recollections between gulps from an oxygen mask. Steve Rowan gave access to his father's diary and sorted through the boxes of old copies of *Caveant Hostes* magazines stored in his garage. A former National Service artilleryman, he helped reconcile the coordinates recorded in the 2/15th war diaries with Captain Smith's Italian 'Tobruch Defences' map. John Neal, an expert on 2/15th history and military matters in general, provided invaluable advice and anecdotal evidence, as did MacKenzie-Smith's brother, Raymond Smith. Greg Snow accessed and processed photographs from Remembrance Club records.

The result was a 100-page manuscript that has been expanded with the inclusion of excerpts from the Rats' diaries and *Caveant Hostes* articles and titled 'Tobruk's Easter Battle 1941: The Forgotten Fifteenth's Date with Rommel's Champion'.

The ultimate endorsement came when the Australian War Memorial acknowledged that the battalion's role in the battle had been overlooked for 70 years. For almost a year, MacKenzie-Smith and Australian War Memorial historian Dr Karl James discussed the issue via email.

In March 2011 MacKenzie-Smith sent a copy of his manuscript to James and soon afterwards received the message that he and everyone associated with the 2/15th Battalion had been hoping for: 'From now on no book on the Tobruk Siege can be written without incorporating the substance of this report.'

Which raises the question: how could history have been so wrong for so long?

'I don't think it was malicious, but I acknowledge that the members of the 2/15th had a genuine grievance,' Dr James says. 'One of the problems was that the Battalion's own war diaries were vague. One of the earliest books written about Tobruk was by Chester Wilmot. He interviewed the company commander of 2/17th Battalion, who described the action in which Jack Edmondson got his VC. Wilmot concentrated on that part of the story and he was so influential that his description set the agenda. Everyone writing about Tobruk for the past 70 years has followed his lead. I was impressed by John MacKenzie-Smith's research. It is good to see the children of veterans telling their parents' stories. The World War II generation is fading away, and if we don't grab their stories now they will fade away with them.'

MacKenzie-Smith donated the bloodied swastika flag that started it all to the Australian War Memorial. It was the centrepiece of the Memorial's display to mark the 70th anniversary of the Siege of Tobruk, with a caption that for the first time publicly stated the 2/15th Battalion's correct position on Easter Monday.

WHEN WILLIAM YATES WAS ten years old, his father gave him his campaign medals and the Military Cross he had been awarded for his actions on Easter Monday. William swapped the campaign medals with another boy for some marbles and never got them back, but he still has the Military Cross and its official citation. If military historians and authors had delved into the Australian War Memorial records and accessed that citation, they would have found exactly what 9 Platoon, A Company of the 2/15th Battalion did during the Easter Battle.

Better still, they could have asked Jack Anning. Now 92 and living in a nursing home in the northern Brisbane suburb of Toombul, Jack's body might be bowed — he uses a wheeled walking frame, the result of a horse falling on him 40 years ago — but his mind is every bit as sharp as the day in 1940 when he walked off the family property at Charters Towers and joined up in Townsville. He remembers names, faces, dates and emotions with crystal clarity. Most of all he remembers that Easter Monday morning.

> I happened to be on guard with a couple of other fellas and just as daylight was breaking I saw the tanks about 3000 yards [2700 metres] away and the troops behind. I was watching straight up the road as clearly as anything, because that's the way of the desert. You can see a long way. I warned the fellas behind me that were asleep. I could see the troops, quite small, come in behind the tanks, form up and come towards

us. There were about 30 tanks and some men riding on them, some trotting along behind. We were in front of the guns, about 150-200 yards. We weren't worried. We were fully trained and we were tired of being chased from pillar to post.

Half an hour after that was when the planes came in, ours and theirs. I remember particularly a plane was shot down and I jumped up in excitement and a fella called out, 'Hey mate, sit down, sit down. He's one of ours.' One tank came up very close to us. It was hit by the artillery and went skewiff and cut across our trench and spun away. We were shooting at the tanks, trying to put one through the slit the driver looks through. Then the tanks started turning around and heading back. Some of the German soldiers took shelter in this trench, which was about waist-deep.

Sergeant Major Robinson was the first to discover them and Greig Smith ordered us down to pick them up and we went down with great excitement, I can tell you. Old Greig told me to leave my rifle and he gave me a Tommy gun and told me to keep an eye on Lieutenant Yates.

We thought there was a just a few of them but when we got down there they started firing, and there was about a hundred of them. That's when I wondered how we were going to get out of it. We started to run out of ammo and Yates told me to conserve my fire. Then Greig sent down two Bren carriers and some mortars. That's who saved the day, because there was still plenty of fight in those Germans. The carriers came one at each end of the trench and did a little bit of machine-gunning. Once the mortar shells started to fall amongst them, they were finished.

The German officer got out of the trench and spoke to Robinson but he wouldn't surrender to him because he wasn't an officer, so Robinson sent him over to Lieutenant Yates. I knew he was some sort of boss because he was slinging his weight around a fair bit but I don't think he had any identification on him. He wanted to organise a surrender; make sure his men were going to be looked after. Then he got back in the trench and started shooting and that's when Yates shot him.

I saw him shot, and I say that man committed suicide. He had no intention of surrendering. It was his second-in-charge who surrendered. That's the fella who put the flag on him. He asked Yates could he kneel down and say goodbye to his friend and he was crying away while he was doing it. Yates naturally said yes, and when he was satisfied the fella was dead he got up and handed over to Yates. Some of the men started to cry.

I couldn't tell you who took the flag off him. All I know was it was taken off. I didn't think too much more about it after that. We had Tobruk and El Alamein and New Guinea and then that was it. I went back to the bush. I think I came down to one

> or two reunions but apart from that my war was over. We won, as far as I was concerned.
>
> I didn't hear any inference that maybe we weren't being treated fairly. I thought we were all soldiers and we all took part in it and won it. It was only when I retired to Brisbane around 1969 that I started to hear talk that we weren't getting credit for what we'd done. I'd never worried about it before. That's not why we were there. When we were fighting we never even thought about acknowledgement or medals.

He opens the storage box under the seat of his walker and pulls out an envelope from which he extracts two small sheets of paper. The pages are yellowed and the writing faded, but the words are clearly legible. It is a letter from Lieutenant Ron Yates to Jack Anning's father. Dated October 9, 1943, it reads: 'Dear Mr Anning. I thought you would like to know that Young Jack has been doing a wonderful job ... '

He grips the handles of the walker and smiles. 'That's my medal,' he says.

Chapter 13

THE BRIDAL TRAIN

A telegram arrived today
It's time to catch The Monterey
'Cause the man I wed, he waits for me
And a daughter that he's yet to see
US Navy beamed its message
Will deliver brides on a one-way passage
It made big news across the nation
The bridal train leaves
from Perth station
(*Bridal Train*, The Waifs)

IT IS OCTOBER 1942 and 17-year-old Joan Staines of New Farm is sitting in a Brisbane cinema beside her date, US Army Sergeant Bill Bentson, 24, of Salem, Oregon, watching a new film called *Bambi*.

It's the part where Bambi is looking for his mum, who has been shot by a hunter. 'Mother? Mother?' says the little deer. 'Where is she?'

A gruff male voice pipes up from the back stalls: 'She's run off with a bloody Yank.'

Around the same time, in Melbourne, 18-year-old Ruth Dowsett is stepping onto the ice at the St Moritz skating arena. Halfway around the rink, her legs become entangled with those of Bill Frost, a 23-year-old soldier with the US Army's 147th Field Artillery, and down he goes. He raises his arms and asks in a South Dakota drawl, 'Well, aren't you going to help me up?'

And in Perth, 23-year-old Betty Denton comes home to the house where she is boarding to find a group of US sailors playing cards at the dining table. Petty Officer Bob Kane, 26, from New Jersey, looks up from his hand and says with a sparkle in his eye, 'And where did this little flower spring from?'

Joan, Ruth and Betty have never met, but they have something in common. They are part of a fraternity that has only recently become recognised and acknowledged for its part in a colourful chapter of Australian history.

They are war brides.

Between 1942 and 1945, up to 1 million American servicemen passed through Australia. With General Douglas MacArthur's Pacific Command headquarters in Brisbane, many were based in Queensland.

As Joan Staines found that night at the movies, the Americans weren't always welcomed with open arms. With the majority of young Australian men fighting overseas, and local girls swept away by the glamorous Americans in their well-cut uniforms, the Yanks were said to be 'over-paid, over-sexed and over here'. The animosity between Americans and locals spilled over in November 1942, leading to two days of riots known as the 'Battle of Brisbane'. But nothing could stop the course of true love.

There is no official figure of how many Australian women married American servicemen and followed them to the US after the war, but it is believed to be as many as 15,000. Some stayed, others returned to Australia with their husbands, or alone, and a few landed in the States only to find they'd been duped by men who already had wives and families.

Macquarie University Department of Modern History researcher Robyn Arrowsmith spent more than seven years working on a PhD thesis about Australian war brides who stayed in the US. She sent questionnaires to about 150, and interviewed 60.

'They've been neglected, really, but they are an important part of Australia's social history,' Arrowsmith says. 'They were fantastic ambassadors, and they were very courageous. At a time when people didn't travel very much, they left everything behind and set off to an alien land.

'Some had very good lives, some had ordinary lives, and some struggled. So little was known about them and their experiences, but they all had stories.'

Stories that only now are being shared and recorded. Joan and Bill Bentson are the couple featured in the War Brides component of the exhibition *Travelling For Love*, at the Queensland State Library. Ruth Frost has self-published a memoir, *Pavlovas to Popcorn*, and Betty Kane's story has been immortalised in song.

OF COURSE, 65-ODD years ago they had no idea of how their stories would turn out. All they knew was that they had fallen in love and nothing was going to stop them marrying their Yank.

'My parents liked Bill, but of course I was young and they thought if we got married I'd go overseas and they'd never see me again,' says Joan. 'But I said if I had to wait 10 years I'd still marry him, so they said all right. My grandmother, my mother and I all put our coupons together and bought my wedding dress.'

Joan and Bill, who now live in Toowong in Brisbane's west, met at a Red Cross dance at Brisbane Town Hall in September 1942.

'I worked at Roma Street markets on the clerical staff. The Red Cross was always looking for girls to go to the dances and one day they came to the markets and asked,' she says. 'It was a Sunday night and it was the first dance I'd ever been to. I went with a friend. Bill came up and asked me to dance. I thought, "He looks all right." He was the first and only boyfriend I ever had.'

What drew him to her? Was it love at first sight across a crowded dance floor?

Sitting beside Joan, 82, on the lounge, 89-year-old Bill quips, 'I looked at the girls and went eeny, meeny, miney, moe.'

'He took me home and asked if he could kiss me

goodnight,' says Joan. 'I said, "No, my mother's probably looking." And she was.'

They were married at St Michael's and All Angels Church at New Farm four and a half months later, just before Bill, who was working on General MacArthur's staff, headed to New Guinea.

'One of my mother's friends said it would never last,' says Joan. 'And here we are 65 years later.'

At war's end Bill was shipped home and in May 1946 Joan received the telegram she had been waiting for. The US government had arranged passage to San Francisco on the liner SS *Mariposa*, departing from Hamilton Wharf. Other ships were leaving from Sydney. A train, now on display at the Bassendean Railway Museum in Western Australia, was dispatched from Perth to Sydney, picking up brides and babies along the way.

From west to east the young girls came
All aboard the bridal train
It was a farewell crossing of her land
She's gone to meet her sailor man
No time for sad goodbyes,
She held her mother as she cried
And then waited there in the Freo rain
To climb aboard the bridal train.

'It was very emotional — there were a lot of tears,' Joan says of boarding the *Mariposa*. 'One girl walked off the ship before it sailed. I was sad but I also thought I was going on a great adventure. I'd written to Bill's mother for almost four years, so I thought I knew her.'

She was the first off the ship when it docked and Bill, who caught the overnight train from Oregon, was there to greet her with open arms.

'As he gave me a hug a huge roar came from everyone on the ship.'

Joan and Bill lived in Oregon for the next 18 years. He worked as a teacher and they had three children: Debra, Brad and Suzanne. In 1964 the family came to Australia — on the *Mariposa* — for what they thought would be a two-year stay. They never went back.

'Missing my family was the hardest part,' Joan says of her time in the US, 'but I enjoyed the rest of it. It was all a wonderful experience. When we were on the ship going over I suppose we were all wondering what lay ahead.'

RUTH FROST (NÉE DOWSETT) was never one to sit back and wait to see where life would take her. She preferred to lead the way.

> I was working in an insurance office during the day and performing in an entertainment unit at night and weekends. We used to entertain the troops in the camps and hospitals, and occasionally on a Sunday night we'd put on a big show at a theatre in the city.
>
> I was in the chorus but I loused it up because I hated routine. Even when I danced I'd ad-lib. I'd be talking to the boys under the footlights. The girls

complained and management said they'd kick me out. I said, 'You couldn't get on without me.' Then I found out they could.

One Saturday night my friend Marj and I couldn't decide whether to go dancing or ice-skating. We tossed a coin and ice-skating won. The St Moritz at St Kilda. That's a lot of saints in there, isn't it, and I wasn't one of them. That was when I met Bill. He still maintains I tripped him on purpose. I always say I tripped him on the ice rink and I've been keeping him on ice ever since. When I helped him up we had a hot chocolate and then he took Marj and me home in a taxi. We'd never been in a taxi before.

Bill was sitting in the middle and said he'd bought a movie camera and wanted to see Melbourne. He said, 'So, would you like to show me around?

And I was thinking, *Pick me, pick me …*

And he said, '… Ruth?'

With Bill soon shipped out to New Guinea, they were together just 31 days over the next three years.

They were married in Melbourne on May 25, 1945, and honeymooned at the Healesville Hotel before Bill returned to active service and was demobbed in the US. Just over a year later Ruth caught the bridal train at Flinders Street Station.

'There were people crying and sobbing everywhere,' she says. 'Parents were running along the platform after the train.'

While Joan Bentson says her passage to the US on the SS *Mariposa* was 'comfortable', Ruth's experience on the army transport ship the SS *David C Shanks* out of Sydney was anything but.

> We sailed Friday and the first Sunday out we hit the backwash of a tidal wave. You've never seen so many green people in your life. There were 300 war brides and 106 babies on that ship and when those seas hit there were only about 12 of us still standing. The nursing mothers were lying around so sick and the poor babies were screaming, they were that hungry. We were grabbing babies and putting them on the mothers. We didn't know who belonged to who, we were just sticking babies on boobs.

With Bill completing an apprenticeship as a linotype operator, he and Ruth lived in the US mid-west, the last 11 years in Manson, Iowa.

In 1960, at Bill's urging, they brought their four children back to Australia. For 23 years Bill worked at the *Hamilton Spectator* newspaper in southwest Victoria. They now live with youngest daughter Janis, 54, at Mooloolaba on Queensland's Sunshine Coast. Ruth, 84, is as feisty as ever and Bill, 89, has lost much of his hearing but not his sense of humour. He greets visitors in a T-shirt that reads, 'I'm not deaf. I'm just not listening.'

BETTY KANE (NÉE DENTON) was one of eight children from Wagin, a small town in the south of Western Australia. Her father died when she was two and her mother never remarried. At 20, she moved to Perth. She was 23 and Bob Kane was 26 when they met.

He was on a submarine tender stationed in Fremantle. They married in September 1944 and Bob sailed to the Pacific three months later. Betty moved back to Wagin to await the birth of their child.

The telegram arrived on February 7, 1946. The bridal train was leaving from Perth Station and Betty had two days to borrow a suitcase and get herself and daughter Susan ready for a new life in America.

She was holding her future in her hand
The faded photo of a man
Catch a sailor if you can
The war bride leaves her southern land

They sailed to San Francisco on the SS *Monterey* and caught the train to Grand Junction, Colorado, where Bob had been discharged from the Navy. He was waiting on the platform. It was after midnight, Susan's first birthday. Betty handed Bob the daughter he had never seen. He handed her a bunch of flowers.

They lived in the US for a few years, during which time their son Peter was born, then moved to Albany, Western Australia. Bob worked as a carpenter and the couple had two more daughters, Pattie and Helen. Pattie had three daughters, two of whom, Donna and Vikki, formed the internationally acclaimed band The Waifs.

Bob died in 2001; Betty, 87, still lives in Albany. The story of their meeting over a game of cards, the crossing from Perth to Sydney on the bridal train, and the uncertainty with which Betty and the other war brides faced the future became part of family folklore.

'I had heard Grandma tell the story over the years,' says Vikki, 'but one day I asked her to tell it in detail and took notes, which I then carried around with me for a few years. One day, six months into a Waifs tour of the US, I was feeling homesick. We were in Whitefish, Montana, and I was walking through the suburbs and the song just started to come to me. First the chorus, and then I went back to the notes and it was already written there, melody and words.

'I was particularly struck by the image of a train full of young women speeding across Australia away from their lives,' she continues. 'Grandma told me that US Red Cross nurses were on the train to help with the children, and she remembered them being happy and vivacious and singing the song, *The Story of a Starry Night*. She said it helped to hear their joy and excitement.'

This is the story of the starry nights
Through desert plains and city lights
Through burning sun and driving rain
She wept aboard the bridal train

In December 2006, Vikki's song, *Bridal Train*, released by The Waifs, became the first non-American song to win the annual USA Songwriting Competition.

On Anzac Day 2007, the Australian Embassy in Washington, DC, held a reception for war brides and their families. It was their first official recognition.

'It was a very emotional night,' says Robyn Arrowsmith. 'There were 95 war brides, all in their 80s. They had travelled from all over the States and brought their families with them. There were 300 to 400 people there. The Embassy put on a wonderful event for them and they were all up the next morning for the Dawn Service.'

As the women left the reception they were given a DVD of *Bridal Train*. It was a song written to tell the story of one of them, but it told the story of them all.

All the girls around Australia
Married to a Yankee sailor
Your fare is paid across the sea
To the home of the brave and the land of the free

Bridal Train was written by Vikki Thorn and is performed by The Waifs. Lyrics used courtesy of Three Little Fish Pty Ltd.

Chapter 14

GUESTS OF HONOUR

EVERY ANZAC DAY THEY come, the little children in their green school pullovers, clutching their posies. Each stands behind one of the 143 gravestones and, at the appointed time, they solemnly place their flowers on a Digger's grave. They sing the English, New Zealand and Australian national anthems and a Year 6 student is chosen to recite The Ode. 'At the going down of the sun, we will remember them …'

It's been this way since the end of the Great War: the children placing their flowers and paying their respects. And their children, and their children.

It's part of growing up here, and of growing old.

With a few variations the latest ceremony, being held on Friday afternoon, will be no doubt similar to those at hundreds, maybe thousands, of primary schools across Australia, but this Anzac Day commemoration isn't in

Sydney or Brisbane or Adelaide. It's at Sutton Veny, two hours' drive from London and 143 lifetimes from Australia.

The graves belong to 141 Aussie men and two Aussie women who came to England to do their bit for King and country in the so-called Great War of 1914–18. Some died of wounds inflicted on the battlefields of France. Others, even more unfairly, if tragedy can ever be quantified, were all but over their wounds and just days away from heading home when they were felled and killed by the influenza pandemic sweeping Europe.

It's the last day of term when I visit Sutton Veny Public School and walk through the graveyard of St John's Church next door. The final bell has just gone and the voices of excited children heading off on holidays waft across the playground as I read the names, dates and ages on the headstones. Most are in the striking style of the Commonwealth War Graves Commission, simply carved on white granite. Others, larger and more ornate, were ordered and paid for by heartbroken parents left grieving on the other side of the world. Their inscriptions, from loving mothers and shattered fathers, make for emotional reading, but sometimes it is not words carved in stone that make the greatest impact. As I pause at one soldier's last resting place, I can't help but say his age out loud. 'Sixteen,' I whisper, shaking my head.

It is the dates, too, that tell the story. The specific geographic origin of the Spanish influenza pandemic of 1918 is still unknown. One theory was that it started in a troop-staging centre in France in January 1918, with the spread of the virus throughout Europe and elsewhere aided by the movement of infected soldiers. Though it was a global tragedy, it became known as the 'Spanish flu' because stories of a heavy death toll in Spain emerged at a time when other European countries had wartime news blackouts in place. Estimates of deaths caused by the disease range between 50 and 100 million over two years. It travelled to the United States, Australia and the Pacific Islands. And Sutton Veny.

TO LOOK AT SUTTON VENY today, it's hard to believe the tentacles of warfare could have reached this far. The description 'nestled' in the English countryside doesn't do justice to the town's rustic isolation. 'Enveloped' comes closer. In fact, if not for the steeple of St John's Church rising above verdant trees, it would be easy to miss seeing Sutton Veny altogether, and to drive straight past.

Perhaps it's why the War Office chose to build an army camp here early in the war. It's so quiet that the soldiers would have had little to do except concentrate on their training. With the bright lights of Bristol a good 48km up the road, they were unlikely to get themselves into too much trouble.

In 1916, with casualties from France mounting by the day, a 938-bed hospital was added to cater primarily for Australian wounded. Greenhill House, a stately home nearby owned by the Walker family of Walker's

Whisky fame, was requisitioned by the YMCA to use as a convalescence centre. Patients considered well enough to head home or back to the front were given a souvenir booklet of postcards showing Aussie soldiers playing chess or football, or even boxing on the manicured lawns of Greenhill, but those in the pictures were the lucky ones.

St John's Church contains a small Anzac chapel in honour of the Aussies buried outside. On the wall near the altar — along with the handwritten names of the dead, a slouch hat insignia badge unearthed and donated by a resident 20 years ago, and an Australian flag — is a framed page from an old magazine. Headed *From Bonnets to Tin Helmets*, it features a photo of nurses working at Sutton Veny Military Hospital and excerpts from a letter written by one of the nursing sisters, describing her patients:

> They are all going back to Australia and are all just shattered wrecks, really ... all my boys are either winged or legs off, shoulders blown away, big head wounds ... some have ONLY had 12 operations.

The nursing sister writes that one of her colleagues is getting married, and her 'one-legged' patients are planning to make an archway with their crutches.

The letter is dated January 28, 1918. The flu arrives in Sutton Veny seven months later.

Judging by the dates on the headstones, the pandemic took hold late in August and was unstoppable by October. One death a week becomes two, then three and four, then five or six a day. October 26 is the worst. Row after row of gravestones record that awful date. Four days later, Matron Jean Walker of the Australian Army Nursing Service succumbs, age 39. She is buried alongside fellow nurses Katie Bolger (a member of the British Nursing Service) and Fanny Tyson, surrounded by 'their boys'.

ON THE OTHER SIDE of the wall separating the school from the church, the children are climbing into their bus or being collected by parents. Waving them off on their Easter break is teaching assistant Nicky Barnard, 51, who for almost 15 years has been Sutton Veny Public School's keeper of the Anzac flame.

The first Anzac Day commemorations were held on 25 April 1916, with services and ceremonies in Australia, a march through London, and a 'sports day' at the Australian camp in Egypt. After the war, Anzac Day became a national day of remembrance for the more than 60,000 Australians who died during the conflict, and the people of Sutton Veny also instituted an annual commemoration of the sacrifice made by the Aussies who came to their village and never made it back home. A service is held each year in St John's Church on the Sunday before April 25. The school holds its own ceremony on the day itself. It is the responsibility of a teacher at the school to organise the event.

Barnard was handed the job in 2000. She didn't need to be told how much the commemoration meant to the school or the village. Her mother had grown up in the town and taken part in Anzac Day ceremonies as a child. But even with that family link, Barnard had no idea how much the story of the Anzacs would touch her and become such a big part of her life.

'There was an Australian liaison officer named Captain Craig Shortt who was stationed over here,' she says. 'He helped us with the service each year and was so interested in the history of it that he got me interested. I wanted to make it so it wasn't just something that the children did every year because we always had. I wanted it to mean something.'

We're seated in a staff area that has become known as 'The Anzac Room'. One entire wall is taken up with a huge noticeboard papered with copies of yellowed photographs of World War I Diggers in uniform. There is an Australian flag presented by the House of Representatives, a frame containing a soldier's diary, and a copy of the Greenhill House souvenir booklet. Most of all, there are folders filled with letters from the descendants of long-dead Anzacs, many of whom lie in their graves 50 metres away on the other side of a low stone wall.

Barnard knows them all, knows their stories and in which book to find them. 'Angus Cameron McPherson,' she says, sliding one from its cover ...

> From Manly, Sydney. He had four kids. Got sick on the boat coming over. Never even got to the war.
>
> Then there's this one, Sergeant Bertie Richard Kerr. He survived the war and stayed in England.

> He died in 1958 and asked to be buried with his mates at Sutton Veny, but because he hadn't died in uniform he couldn't be part of the war graves. They buried him nearby on the other side of a little wall and then some years later that wall was taken down, so he is with them now after all.

'Here's another one for you. A Queenslander.' She passes me a letter from Lorraine Taylor and family of Buddina, on the Sunshine Coast. Mrs Taylor's grandfather, Charles Hearn, died at Sutton Veny of influenza on May 1, 1919, the day he was supposed to sail home. Her daughter, Robyn, visited his grave in 2002.

'I would just like to say how much the whole family appreciates the care taken of these gravesites of soldiers who came from the other side of the world,' the letter reads. 'Although most loved ones or relatives have not, and may never visit these graves, it's comforting to know thousands of kilometres away caring people, like yourselves, are tending to them.'

Each year, about 20 Australians call into the school and visit the graves. Those who attend the Anzac Day ceremony never leave unmoved. Adelaide woman Julie Reece was so touched that she wrote and self-published a superb book tracing the wartime experiences of her uncle. Called *Jimmy's Anzac Pilgrimage*, it's subtitled 'A tribute to Private James Martin Neagle and the children of Sutton Veny who have never forgotten the Anzacs'.

Barnard reaches into a drawer and pulls out a handful of red silk poppies. 'Three hundred and fifty of these arrived last week from Geraldton RSL in Western Australia,' she says. 'One of their members came to our service in 2005, and when he got home he put it to the RSL that they do something to help us. Now these poppies arrive every year.'

RETIRED BRISBANE SCHOOLTEACHER Neville Biggs, 72, and his son Geoff, 44, will visit Sutton Veny later in the year. It is part of a journey following in the wartime footsteps of Neville's father, Ted, who served with the AIF 37th Battalion on the Western Front. After seeing action at Albert and being involved in the liberation of Villers-Bretonneux, Ted's battalion fought eastwards along the Somme River through Proyart, Péronne and Mont St Quentin. In early October 1918, on the last day that the 37th would see action in the war, Ted was gassed near the villages of Le Catelet and Gouy. Repatriated to England, he was blinded for three weeks. After treatment in Kent and Surrey, he was sent to Sutton Veny for rehabilitation before returning to Australia.

'He never spoke much about the war at all,' says Biggs from his home on the Sunshine Coast. 'But he did tell me the story of how he felt unwell and went to his sergeant, who looked at the blisters on his neck and told him he'd been gassed and to get himself to the Regimental Aid Post. On his way he collapsed on the ground, and when he woke up he couldn't see or talk or move. He could hear two Americans walking past and one of them said, 'He's a goner.' But just then Dad dry-retched and they realised he was alive and took him to the RAP. When he got his sight back, he was being treated at Horton Hospital in Surrey, a former mental asylum. The first thing he saw was the bars on the windows and he thought he was in jail. When he got back to Australia he suffered from the effects of the mustard gas for the rest of his life, but he lived until the age of 80.'

Every year, Biggs, a former deputy principal of Ferny Grove State High School in Brisbane's northwest, writes and produces a musical for his local Uniting Church at Maroochydore. Geoff, a video forensic analyst with Australian Federal Police, films it for the cast and parishioners.

'This year I thought I'd do something to mark the centenary of the start of World War I,' says Neville Biggs. 'I didn't want to set it in a battlefield, so I thought I might put it in a repatriation hospital. In my research, I came across the story of Sutton Veny and it reminded me that my father had been there.'

Neville's play is set in Greenhill House on Anzac Day, 1918. It will be performed in June and recorded by Geoff. In September, the two men will visit Sutton Veny and present a copy to the school. No doubt it will be added to the collection of gifts, souvenirs and mementoes that are used each year as part of the children's Anzac education.

'Every child in the school takes part in the Anzac Day service,' says Nicky Barnard, who visited the battlefields of

Gallipoli in 2002. 'This year it will be based on the book, *Anzac Biscuits*, that someone sent us. It's a wonderful book. It tells the story of a soldier in the trenches while his wife and daughter are back in Australia making the biscuits. One of the boys will be wearing a real World War I Australian uniform to play the soldier. It will be beautiful.'

And then, after the play is over, the children will take their places behind the gravestones of Matron Walker and Angus Cameron McPherson and Charles Hearn and 140 of their comrades, and respectfully place their flowers as a Year 6 student from Sutton Veny says the words that have been said here for as long as anyone can remember:

'Lest we forget.'

PUBLISHER'S NOTE

By Geoff Armstrong

STOKE HILL PRESS IS a sports publisher. So why did I decide to publish this collection of war stories? One reason is that Mike Colman is a friend, a good man and a great writer. Over the years, beyond his impressive catalogue of sports writing, he has produced some superb award-winning stories on the subject of war, not so much about the battles as about the soldiers who fought in them and the people back home who were affected by them. They deserve to be collected in one volume.

For me, it resonates deeper than that. Mike's war stories are about ordinary people (if 'ordinary' is the right word), who found themselves in extraordinary circumstances and were asked to do extraordinary things. When he writes of Ray McMillan or Keith Payne, Charlie Blackman or Jack Anning, Cliff Hopgood or Charlie Marshall, he could just as easily, in my mind, have been writing about my grandfather, Francis Edwin Armstrong. It's not that my grandfather's story is similar to any other told in this book — it seems to me that no two war stories are ever the same — but like the heroes Mike meets or remembers in his stories, 'Pa', as we grandchildren called him, also went far beyond the call of duty, and made us all eternally proud of him.

Ted Armstrong was a schoolteacher. He was 21 years old and working in classrooms in around his hometown of Wallsend, about 10km from Newcastle, 160km north of Sydney, a world away from France and England, when he enlisted for the Great War on July 24, 1915. As a kid, he'd been resourceful enough to earn a scholarship to Newcastle High School, 'tenable for three years [and] with a free supply of textbooks'. He joined the 19th Battalion and five months later left Sydney on the HMAT *Suevic*, bound for Egypt. The voyage, he wrote in his diary, was 'monotonous', but such was his sense of adventure in the days and weeks before he was summoned to the trenches, he fell 'in love' with his new surroundings. The area around the Ferry Post at the Suez Canal would, he wrote, 'do me for a long time'.

That, of course, is not how war works. Transferred to the 55th Battalion as part of the reorganisation that followed the Anzacs' Gallipoli campaign, he was promoted to sergeant, shipped to France, and suffered gunshot wounds at the Battle of Fromelles on July 20, 1916, during an exchange he described as 'veritable hell'. More than 5500 Australian soldiers lost their lives — a disaster the renowned military writer Les Carlyon described as 'one of the worst in Australian history, probably *the* worst in terms of the scale of the tragedy and the speed of it, a mere 14 hours'. Among the dead was Lieutenant Berrol Mendelsohn, aged 25, from Bondi in Sydney, a swimmer of some ability who had served at Gallipoli before he joined the 55th. Five months later, Pa wrote to his late comrade's family.

> A friend of mine, Pte A.E. Rodda, received a communication from you, asking for particulars of the death of Lieut. B. Mendelsohn of this Battalion and as I, perhaps better than anyone else, can supply those particulars, I have undertaken to write you, and do so.
>
> Lieut. Mendelsohn was platoon commander of No. 3 Platoon, and I was platoon sergeant of No. 4 Platoon. On July 19, when I reached our trenches with my platoon, Lieut. Mendelsohn, by some chance separated from his own platoon, was near us. He immediately took command of those men near him, and blowing his whistle led the way over the parapet towards the German trenches.
>
> When I reached the German first line of trenches, I found that Lieut. Mendelsohn was not with us, but on reaching the German second line, he was already there. This was about 7pm, July 19.
>
> As soon as I reached this position Lieut. Mendelsohn gave the order to move along the trench and we occupied a portion of the trench previously unoccupied.

> We were together here all night, except for a short period when Lieut. Mendelsohn moved down the trench.
>
> At about 2am on July 20, the Germans counter-attacked heavily and we stood to, to withstand the attack, Lieut. Mendelsohn in command. At about 2:30am, he was shot through the head standing alongside me, whilst urging his men on to greater effort. Death was instantaneous. I was myself wounded shortly after, but have ascertained that Lieut. Mendelsohn's body was not removed from the trench and was probably buried by the Germans.

The remains of Berrol Mendelsohn, a great-uncle of the actor Ben Mendelsohn, would not be formally identified and re-buried until 2010, after archaeologists searched an area at Pheasant Wood, near Fromelles, and discovered mass burial pits that had been missed by grave recovery parties immediately after the war. What might he have achieved, had he survived the war? General Sir William Birdwood no less, commander of the Australian forces on the western front, wrote to Mendelsohn's mother Abigail: 'He was an officer of the highest ideals, very efficient, and his loss has consequently been most severely felt by the battalion, and by his colonel, who regarded him as such a trustworthy friend.'

After rehabilitating in England, my grandfather returned to the frontline two months after the carnage at Fromelles, was promoted to lieutenant, and survived fierce exchanges with the Germans at Amiens, Villers-Bretonneux and Morlancourt. At the end of September 1918, during the 55th's assault on the Hindenburg Line north of Bellicourt, his 'conspicuous gallantry and devotion to duty' earned him a Military Cross, the citation reading in part:

> Early in the attack, the attacking troops were held up by the enemy's extremely heavy machine-gun fire. Lieut. Armstrong, moving amongst the men of his Company, encouraging them by his personal bravery and total disregard of fire, reorganised them for the final rush at the M.G. nests. Jumping out at the head of the Company with the cry 'Come on Lads', his men bravely followed him with a cheer as he struggled forward in the face of the spitting guns. Urged on as they were by this officer, his men pushed home the assault with bayonet and bomb with such determination and dash that the Huns were overpowered before they realised what had happened. The result was that the danger from the enemy's miniature fortress which was holding up all the attacking troops of the Brigade was now removed and the advance of the whole line was allowed to continue unharassed.

Family folklore has it that Pa received his Military Cross personally from King George V. From what my dad told me, my grandfather rarely talked about his war experience, at least not with members of his family. He died on November 29, 1966, when he was 73, and I had just turned 5, so I never had the chance to pry. When he returned to Newcastle in 1919, he told a welcome-home reception that he'd 'been on the move for so long it would require some time to get back to the old order of things'. And he added flatly: 'I, with the others, went away to do my bit, and if you are satisfied that that has been done, and are satisfied to have me back again, I am satisfied to be back.'

With that, he thanked everyone for their 'hearty welcome' and retreated back into civilian life.

Or at least I imagine that might have been the plan. Ted Armstrong returned to teaching, working at various schools in the Hunter region, but he quickly became dispirited with the way many returned soldiers were treated. So began a lifetime of civic service. He became President of the Wallsend branch of what we now know as the Returned Services League (RSL), and in November 1922, the *Newcastle Morning Herald* reported, decided to cancel that year's annual Armistice Day picnic 'on account of the distress existing amongst Diggers'. Instead, he asked everyone to 'contribute to the Red Cross Society for the purpose of aiding their work of relieving the distress'.

Late in 1924, he was appointed to the headmaster position at Baradine Public School in western NSW, and two years later was transferred to nearby Binnaway Public School, where he also took over as president of the RSL, was elected president of the local Progress Association and joined the committee for the Binnaway Show. In 1933, he returned to the Newcastle area, as headmaster at Boolaroo Public School, where he stayed for seven years, until he was called to the city to take charge at Parramatta North Public School.

My grandfather would become a prominent public figure in western Sydney. I'm not sure he ever sought the spotlight, but he'd learned that you could hardly ask others to get involved if you weren't prepared to do so yourself. He joined the Volunteer Defence Corps. He was elected to the committee of Parramatta RSL and was named president in 1945. A year later, he became an alderman on Parramatta Council. His sons went to Parramatta High, and he soon became the secretary of that school's Parents and Citizens Association. He was one of the founders of the Parramatta division of Legacy, was on the committee of Parramatta Rotary and was a member of the City of Parramatta War Memorial Trust. His motivation might have been captured in part by his remarks at an Armistice Day service in 1945: 'Remember not only the men who died. Remember also the men who lived. Help them.'

Nineteen months earlier, Pa had spoken passionately at an Anzac Day 'smoke concert' at Parramatta's Soldiers' Hall:

> To me the real significance of the day we celebrate is not merely to remember the landing at Gallipoli as a glorious operation — which it was — but to

> remember the old Anzacs and the new Anzacs, and to pledge ourselves that the new Anzacs who come back will have a better spin than the old ones did ...

He said that for the participants of the two Great Wars, different days had 'their own significance, their own memories'.

> The day we celebrate is not merely April 25, 1915, but all those other days I have spoken of. We gather together to remember our mates who were left behind, whether in the first war or the second. As we think of them, what do they think of us? They're looking down on us and asking what we have done in their memory.
>
> I don't think they want us to mourn. They want us to remember them, but not to go around with gloomy faces. They want us to remember them as they fell with a smile on their faces. They want us to remember those they have left behind.
>
> After the last war, what happened to the dependants of those who fell? What happened to those men who came back — and wanted work? You know what happened! What's going to happen to the men who come back this time? To the dependants of those who fall in this war?

Ted Armstrong's commitment to his fellow returned soldiers and the wider community never wavered. He shifted from Parramatta North to Blaxcell Street Public School at nearby Granville in 1953, and succeeded in turning his new responsibility from a place many parents tried to avoid into 'one of the greatest schools in the state'. That's how Maurice de Ferranti, the District Inspector for the Education Department, described it.

'I have always believed in handing responsibility to staff, young or old, and treating members of staffs as members of a co-operative organisation,' Pa said.

Even after he retired at the end of 1958 he continued to work as a maths teacher until his death — in part because, as the city's population boomed, good teachers were hard to find, but also because, simply put, he loved it.

He was one of the 'lucky' soldiers in that he lived a long life and he had the chance to let his experiences — on the battlefields of France and later as a teacher and good citizen — shape his life and character.

In 1953, as president of the Parramatta RSL, he was asked to speak at a function organised to honour a 22-year-old cricketer from the local Cumberland club named Richie Benaud who had just been chosen to tour Britain with the Australian cricket team. Cumberland (now known as Parramatta) had previously provided three Test men — Gerry Hazlitt, Frank Iredale and Bill Howell — but they had learned their cricket elsewhere before joining the club as established players. Benaud was the first cricketer from Parramatta to wear the baggy green. It is impossible to underestimate the pride the district felt for their new hero.

'Richie, your skill has gained you the honour of an English tour,' said Ted Armstrong, the RSL president, headmaster, the soldier who had received his Military Cross from the King. 'It is unique, because you are the first local boy to achieve this distinction. You will associate with the highest in the land …

'I hope that you never lose the common touch.'

That was my grandfather's advice. That's what mattered to him, an 'ordinary' person who did extraordinary things. I can imagine Ray McMillan and many of the other leading characters you have read about in this book offering a similar suggestion. I'm sure there are many families who have forebears whose experiences and ordeals fighting for Australia shaped them in a similar way. We rely on exceptional writers such as Mike Colman to tell their stories, and to tell them well, so we can remember them and learn from them.

That's why I was so keen to publish this book.